CYCLES IN THE SKY
Second Edition

Michael C. LoPresto
Henry Ford Community College
Dearborn, MI

KENDALL/HUNT PUBLISHING COMPANY
4050 Westmark Drive   Dubuque, Iowa 52002

Cover image ©2004 Photo Disc, Inc.

Portrait Illustrations by Colin Ferguson

Copyright 2002, 2004 by Kendall/Hunt Publishing Company

ISBN 0-7575-1053-1

Printed in the United States of America

10 9 8 7 6 5 4 3 2 1

DEDICATION

To Jan;
The brightest star in my sky
And our satellites;
Sarah, Emily, and Sam

CONTENTS

ACKNOWLEDGMENTS

I thank the Regional, Acquisitions and Copy Editors at Kendall Hunt for encouraging me to write for and publish with them, guidance on technical details, and editing the manuscript. I also thank my former Astronomy student Shadden Sion for her review of the manuscript of the first edition and my colleague Steven Murrell for his review of the second. Mostly I thank the hundreds of students who have taken introductory astronomy with me since 1990. Without them as an audience the ideas for this book or its revision would never have developed. A special thanks to the students who pointed out errors in the first edition.

The portraits were prepared by Colin Ferguson, Assistant Professor of Art at John Tyler Community College in Midlothian, Virginia, a close friend since the third grade and among many other things, a college roommate and groomsman in my wedding.

I thank the UC Berkley Webmaster for permission to use the figures on extra-solar planets and the Director of the Edinboro University-Maize Sunfire Observatory for permission to use images taken there.

I also thank Dr. David J. Krause, retired Instructor of Astronomy and Geology at HFCC, for the almost 12 years we worked together. His influence has played a large role in shaping the way I teach astronomy.

Last but not least, I thank my wife Jan and children Sarah, Emily and Sam for their patience with a husband and father who is just a busy as his wife and kids. I thank my father, Dr. James C. LoPresto of Edinboro University of Pennsylvania, who was my first and still best example of an astronomer and a scientist; my mother, Meilute LoPresto, for all her love and support over the years and her parents, Gerry and Nellie Juskenas. Without their help I may not have been able to reach my educational and professional goals. This is also true of my father's parents Charles and Jessie LoPresto, whose memory still lives on in my heart and mind.

INTRODUCTION

I have taught introductory astronomy at Henry Ford Community College since 1990 and have used a number of different textbooks. For the most part, I have felt that they all have been fairly good in their coverage of the solar system, stars, and galaxies and the universe, the subjects often covered in the second, third and fourth major sections of a traditional course. The first section, which is usually devoted to motions in the sky and the history of astronomy, is the one with which I have rarely been satisfied. I have liked the way some texts handle it better than others, but I have never been able to find one that covers this very important material quite the way I would like.

What is it that I would like? An often-stated goal of many general education science courses for non-science majors, including introductory astronomy, is to use the specific discipline to expose students to the process of science. There is a wonderful opportunity to do just this at the beginning of an introductory astronomy course that is often missed by many textbooks and instructors. It involves first describing the observable motions of the stars, the Sun, the Moon and the planets and developing possible explanations for them, then, by using the history of astronomy, showing how these theories evolved. This provides not only a lesson in the process of science, but it also shows how the scientific method was developed.

Too often, I see textbooks list the steps of the scientific method like cookbook instructions then gloss over the sky's motions and history on the way to "more important" material. Also, even if the motions are covered in depth, many texts immediately jump to the rotation and revolution of the Earth to explain them without any discussion of how these ideas were first developed.[1]

I offer <u>Cycles in the Sky</u> as my solution to these problems. "<u>Cycles</u>" as it has come to be known, is designed to stand alone as a more in-depth look at the motions in the sky than is normally found at the introductory level, but it can also serve as a supplement to the traditional introductory astronomy text. "<u>Cycles</u>" has been used in this fashion at HFCC in the early parts of the course, when sky motions and history are being covered.

Most of the material in this book comes from my years of lecturing on these topics, but there is much more detail than I have usually had time to cover. Also, in light of the recent move toward more active and inquiry-based learning in introductory astronomy, activities that I have used over the years are included. They can be done during class time or assigned as homework to help the students understand the concepts.

This second edition reflects what are hopefully improvements that have come about from what was learned while the first edition of "<u>Cycles</u>" was used at HFCC by myself and several colleagues with approximately 800 students of introductory astronomy in the classroom over the last two academic years.

Michael C. LoPresto,
Dearborn, MI, 2004

[1]For more details on my views on teaching introductory Astronomy, see: M. LoPresto, *Teaching the Scientific Method in Introductory Astronomy*, **Astronomy Education Review**, Vol. 2, Issue 2, (aer.noao.edu).

Chapter One

MOTIONS
OF THE STARS

OUR VIEW OF THE SKY

From our perspective, the Earth appears to be a disk that is centered on the observer and extends about the same distance in all directions until it ends at the **horizon**, the line where the Earth ends and the sky begins. The sky appears to be a hemispherical shell or dome that also seems to center on the observer. The point directly overhead is called the **zenith**. The points due north, south, east and west on the horizon are known as the **cardinal points**. The line that extends from due north to due south passing through the zenith and dividing the sky into an eastern and a western half is called the **meridian**.

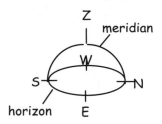

Figure 1.1 The sky from an observer's perspective

HORIZON COORDINATES

Two coordinates specify the position of an object in the sky. **Altitude** is an object's angle above the horizon. The altitude of an object can range from 0° on the horizon to 90° at the zenith. **Azimuth** specifies a point on the horizon that an object is directly above. Due north is

1

defined as azimuth 0°, due east is 90°, due south is 180°, and due west is 270°. Altitude and azimuth together are known as **horizon coordinates**.

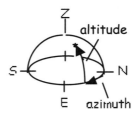

Figure 1.2 Horizon coordinates

OBSERVING THE MOTION OF THE STARS

The night sky is filled with points of light that we call **stars**. If you watch the stars for a while, you will notice that they move. If you note a star's position relative to a landmark, perhaps a tree or a building, and look for the same star about 20 or 30 minutes later, you will notice that its position relative to the landmark has changed. This will be true of almost any star you observe.

Stars do not move individually, they move together as if the whole sky is moving around you. Most stars come from below the horizon, or **rise**, in the East, move across the sky, cross the meridian, then disappear below the horizon, or **set**, in the west. When a star crosses the meridian, it is said to be in **transit**. Transit is the star's highest altitude. The amount of time between two transits of a star is the amount of time for a star to complete one cycle of its motion. This cycle is called a **sidereal day**, a day according to the stars.

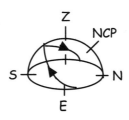

Figure 1.3 The daily path of a star

2

THE NORTH STAR

The paths that the stars appear to follow across the sky are circular. All circles have centers. The point at which the stars' circular paths are centered is called the **North Celestial Pole** (NCP for short, as shown in Figure 1.3). This point is called the North Celestial Pole because it is located on the meridian directly above the point due north on the horizon. There is a star that is located almost exactly at the North Celestial Pole. It is called the Pole Star or **Polaris** in Latin. Since Polaris is very close to the NCP, it has very little apparent motion, which makes it a good marker for the position of the NCP and the direction North. Due to its unique position, Polaris is also often called the **North Star**.

A trick for finding the North Star is to use one of the most recognizable groups of stars in the sky, the **Big Dipper**. The Big Dipper is part of the constellation **Ursa Major**, the Great Bear. **Constellations** are groupings of stars that are used to divide the sky into 88 different regions. Imagine a line pointing out of the Big Dipper's bowl that goes through the two stars at the end of the bowl, Duhbe and Merak. The line will point right to the North Star. The North Star is the end of the handle of the **Little Dipper**, also known as **Ursa Minor**, the Little Bear.

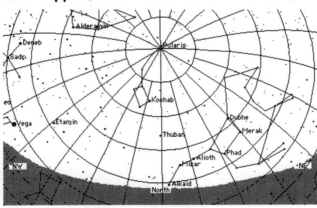

FIGURE 1.4 Use the pointer stars to find the North Star

3

GEOGRAPHIC COORDINATES

Finding Polaris is very important for navigation, not only because it tells you which way is North, but also because it tells you *where* you are. This fact was especially important to navigators of land and sea before the invention of compasses and other navigation equipment. The altitude of Polaris is equal to the **latitude** of the observer. Latitude is the angle of a location north or south from the Earth's **equator**. The equator is the line that divides the Earth into a northern and a southern half. The latitude of the equator is 0°, and the latitude of the North and South Poles are 90°N and 90°S, respectively.

A point on the surface of the Earth can be defined by specifying its latitude and its **longitude**. Longitude is a measurement of how far east or west a location is from the **prime meridian**. The prime meridian is the line of longitude that passes through Greenwich, England and is arbitrarily defined as 0°. Longitudes run from 0° to 180° both east and west, meeting at the **international date line** on the opposite side of the Earth from the prime meridian. Latitude and longitude are known as **geographic coordinates**.

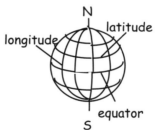

Figure 1.5 Geographic coordinates

It is important to understand that the choice of the equator as 0° latitude is a natural one because it divides the Earth into a northern and a southern half. Any point on the equator is halfway between the North and South Pole. The choice of the location of the prime meridian is arbitrary. Greenwich, England (near London) was chosen because it is the location of the Royal Greenwich Observatory, which was among the most prominent observatories in the world when the decision was being made.

EXPLAINING THE MOTIONS

Now that the motion of the stars has been observed, an important next step is to ask some questions. Why do the stars move the way they do? Why does the altitude of the North Star equal the observer's latitude? To answer these questions, we must attempt to understand the observed motion of the stars. An attempt to explain an observation is known as a **model**.

A model used to explain the motion of the stars is known as the **celestial sphere**. The celestial sphere, shown in Figure 1.6, is an imaginary sphere that represents the sky and is centered on the Earth. All objects in the sky appear to be attached to the celestial sphere. All points on the celestial sphere correspond to the points on the Earth that they are directly above. The **North Celestial Pole**, or NCP, is the point above the Earth's North Pole; the **South Celestial Pole**, or SCP, is directly above the Earth's South Pole; and the **celestial equator**, CE, is the line directly above the earth's equator.

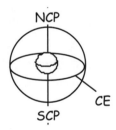

Figure 1.6 The Celestial Sphere

EQUATORIAL COORDINATES

The coordinates on the celestial sphere are called **declination** and **right ascension**. Declination is very similar to latitude. The celestial equator is 0° declination, and the North and South Celestial Poles are +90° and -90°, respectively. Right ascension is similar to longitude.

However, since the celestial sphere appears to move around us once per day, the sky is divided into 24 hours instead of 360°. The declination and right ascension specify the position of any object in the sky. Together they are called **equatorial coordinates**.

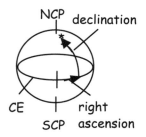

Figure 1.7 Equatorial coordinates

WHY WE SEE WHAT WE SEE

The part of the celestial sphere that an observer will see depends on their location on the Earth. An observer at the Earth's North Pole will look straight up and see the NCP at their local zenith, 90° below their zenith, on their local horizon, they will see the CE. They are seeing the northern half of the celestial sphere. Note that since it is at the zenith, the altitude of the NCP, 90°, is equal to the observer's latitude, which is also 90° at the North Pole. As the sphere appears to rotate once per sidereal day, the stars will all move around the observer in paths parallel to the horizon. Therefore, no stars will rise or set. The stars in the northern half of the celestial sphere will be seen all the time.

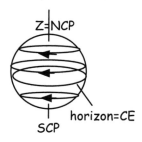

Figure 1.8 The view from the North Pole

6

At the equator, an observer will look up and see the CE going through the zenith. The NCP and SCP will appear on the horizon. From the equator an observer is seeing a "side" half of the celestial sphere. Notice that the altitude of the NCP, now 0°, is again equal to the latitude, which is 0° at the equator. As the sphere appears to rotate, all stars are perpendicular to the horizon; therefore, they will all rise and set. All stars visible from the Earth can be seen at one time or another from the equator.

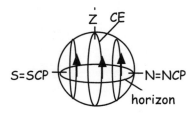

Figure 1.9 The view from the equator

At middle-north latitude (most of the Continental United States, Europe, and Asia; the location of the majority of world's population), the NCP will appear in the northern sky at an altitude equal to the observer's latitude and the CE will be in the southern sky.

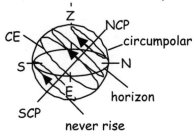

Figure 1.10 The view from middle-north latitude

As they move in circles around the pole, the stars closest to the NCP will never actually set. There will also be stars close to the SCP that will never rise. Stars that never set are referred to as **circumpolar**, meaning "around the pole."

7

At the North Pole, all the stars are circumpolar and stars south of the celestial equator are never seen. As you move south more southern stars become visible and less stars are circumpolar. When you get to the equator none of the stars are circumpolar. As you move into the Southern Hemisphere stars around the SCP become circumpolar and stars near the NCP are not seen. People in the Southern Hemisphere never see Polaris, and there is not a bright star that marks the SCP (there is no "South Star"). At the South Pole all the stars are once again circumpolar, but they are the stars from the southern half of the celestial sphere. If a penguin living near the South Pole and a polar bear from near the North Pole compared star charts they would think each other crazy; the charts would be completely different. Inhabitants of the Southern Hemisphere consider the northern circumpolar stars like those of the Big and Little Dippers just as unfamiliar as Northern Hemisphere inhabitants find the Southern Cross and the Centaur.

The celestial sphere is a global model that accurately predicts what portion of the sky observers from various locations on the Earth will see from their local perspective. When an observer looks directly overhead they define their zenith. Since by definition, the local horizon is 90° from the zenith, the observer will be able to see portions of the celestial sphere up to 90° from the declination at their zenith. Also, as we have seen, this results in observers at any location in the Northern Hemisphere seeing the NCP at an altitude that will be equal to their latitude.

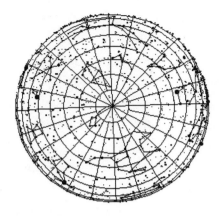

Figure 1.11 Chart of the Northern Sky

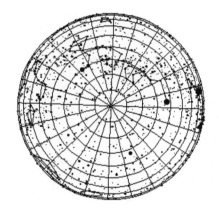

Figure 1.12 Chart of the Southern Sky

Activities

1. Visit a planetarium to observe the motions of the stars discussed in this chapter.
2. Better yet, go outside and see the motions yourself. **Observational Activities** can be found in **Appendix 1**.
3. Complete **Celestial Globe Activity #1-The Stars**, found in **Appendix 3**.

Questions

1. Describe how the position of the Pole Star will change as you move north or south. As you move east or west?

2. If you are at the North Pole, what part of the celestial sphere will you observe at your zenith? At your horizon?

3. If you are at the equator, where should you look for the North Celestial Pole? The South Celestial Pole? The celestial equator?

4. Is there a location on the Earth where you would have a chance to see every star visible from Earth at one time or another? If so, name the location.

5. What location(s) has (have) the most circumpolar stars? The least?

6. If you are traveling in the Northern Hemisphere and see that new stars, ones you haven't seen before, are appearing, in what direction are you traveling? Explain how you know.

7. If it is clear out and you cannot see Polaris, what is something you can definitely say about your position?

8. Identify the following points, lines, or coordinates as either local or global: altitude, azimuth, cardinal points, celestial equator, celestial pole, declination, horizon, latitude, longitude, meridian, right ascension, zenith.

9. If you travel from one location to another, which changes: the altitude and azimuth of a star, the right ascension and declination, both, or neither? Explain your answer.

10. If you observe the sky for several hours, which changes: the altitude and azimuth of a star, the right ascension and declination, both, or neither? Explain your answer.

11. Would Detroit, Michigan or Tucson, Arizona have more circumpolar stars? Why? Which location would have a higher altitude for the Pole Star?

12. If you live in Michigan and call a friend in Arizona to tell them to look for a star that you saw, which coordinate system should you use to tell your friend the location? Explain your answer.

Chapter Two

MOTIONS
OF THE SUN

DAILY MOTION

The daily motion of the Sun is similar to the motion of stars. It rises in the East, increases in altitude until it crosses the meridian, and then decreases in altitude until it sets in the west. The time that the Sun transits the meridian is defined as local **noon**. The amount of time between two "noons" is defined as a **solar day**, a day according to the Sun. This is the 24-hour day of our calendars and clocks. The origin of the time units, hours, minutes, and seconds, have no real physical basis, they are just arbitrary divisions of the day. The second used to be considered 1/60th of 1/60th of 1/24th, or 1/86,400th, of a solar day. Now it is defined in terms of the vibrations of a certain isotope of the cesium atom and also the speed of light. The solar day is actually about 4 minutes longer than the sidereal day. The reason for this discrepancy will become clear once the reasons for the Sun's motion have been explained. The hours before the Sun reaches the meridian are called "*ante meridiem*," AM being the familiar abbreviation for the morning hours. The "*post meridiem*" hours, abbreviated as PM, are after the Sun has crossed the meridian.

TIME ZONES AND DAYLIGHT TIME

The definition of noon being the transit of the Sun is the reason for **time zones**. As the Sun appears to move across the sky from east to west, it will cross the meridian at eastern locations before those in the West. Noon in New York occurs 3 hours before it does in Los Angeles.

The world is divided into 24 time zones, the center of each defining noon as when the Sun is actually on the meridian. If you are east of the center you will see the sun transit before noon, west of the center it will transit after noon. Since Cleveland is west of New York, the center of the eastern time zone, people in Cleveland see transit after noon. Since St. Louis is east of Denver, the center of the Mountain Time zone, people in St. Louis see transit before noon.

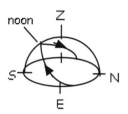

Figure 2.1 The daily motion of the Sun

Daylight savings time, state lines, and politics further complicate things. Daylight time was originally invented to give farmers an extra hour of daylight to work in the fields. Setting clocks an hour later from spring until fall makes the Sun appear to cross the meridian a whole hour later through half of the year.

Despite being much closer to, Chicago, the center of the central time zone, than to New York, Detroit is on eastern time due to the fact that leaders of the automobile industry wanted to start their business day at the same time that the banks opened on Wall Street.

The dividing line between the eastern and central Time zones goes through Indiana. Parts of the state do observe daylight savings time and others do not. This can make traveling through Indiana very interesting.

ANNUAL MOTION

The difference between the motion of the Sun and that of the stars is that, observed from a given location, a star will rise and set at the same azimuths each day and transit the meridian at the same altitude. However, the Sun's path is slightly different each day. If you

12

observe the rising and setting positions and transit altitude of the Sun on a spring day, for instance, you will not notice much change in the next day or two. However, after about a week or more you will notice a difference. The rising and setting positions will move northward along the horizon and the transit altitude will increase. This change will gradually continue until around June 21. At that time the rising and setting positions reach their northernmost points and the transit altitude reaches a maximum. The Sun's path through the sky will be its longest, so we will experience our longest daylight period. This "longest day of the year" is called the **summer solstice**. The term *solstice* means, "Sun standing still." Around the time of the solstice, the Sun's midday altitude does not seem to change much for a week or more.

Since the Sun is high in the sky, its rays strike the ground more directly and heat the ground more efficiently, causing the warm summer weather. When the sun is lower in the sky, the rays are much less direct. The same amount of solar energy is spread out over a larger area, and the ground is heated less efficiently, see Figure 2.2.

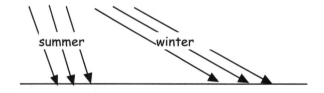

Figure 2.2 The Sun's rays from high and low altitudes

Throughout the summer the Sun's path moves southward and the transit altitude gets lower. Since the path is getting shorter, the daylight periods get shorter. Eventually, by about September 21, the rising and setting positions reach exactly due east and due west and the Sun is up and down for 12 hours each. This is known as the **autumnal equinox**, from which the word autumn originates. *Equinox* means "equal night." Since the Sun's altitude is getting lower, the weather, as we expect in the fall, becomes cooler.

The Sun will continue to move south and the transit altitude will continue to decrease until about December 21. The day with the shortest path and lowest midday altitude is called the **winter solstice**. After this solstice, the Sun's rising and setting positions will start to move back toward the north and its midday altitude will increase. When the rising and setting positions return to due east and west, the transit altitude will be the same as it was on the autumnal equinox. This is the **vernal equinox**, the first day of spring.

Eventually the Sun's path will return to being identical to the one first observed. This longer cycle of the Sun is called the **tropical year**. So, like the day, the year is also a natural cycle of the Sun. The tropical year is approximately 365 1/4 solar days. The extra 1/4-day is the reason that we have leap year. Our calendar is 365 days, so every 4 years a day is added at the end of February to keep our calendar in sync with the Sun. Our calendar will be discussed in more detail in **Chapter 5**.

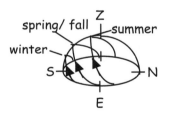

Figure 2.3 The Sun's path on the Solstices and Equinoxes

The passage of the tropical year can be observed either by watching the change in the Sun's rising or setting positions on the horizon or the changes in its midday altitude. In the example discussed above, day after day the Sun's rising position would be observed to move toward a northernmost point, then start moving south, passing due east and eventually reach a southernmost point. It would then start moving north, again passing due east, and eventually return to the point of the initial observation. These observations could also be made with the setting positions of the Sun in the west.

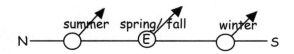

Figure 2.4 The rising position of the sun changing throughout the tropical year

If you observe the Sun at the same time every day, its cycle can be seen in terms of its changing altitude. The reason that the path seen in Figure 2.5, called the **analemma,** is a figure-eight is because the solar day is not exactly 24 hours; sometimes it is a little more, other times it is a little less. 24 hours is the mean or average solar day. So, as seen in Figure 2.5, observing at noon every day from Chicago (at the center of a time zone) will find the Sun on the meridian some days, but usually either a little bit in front of or beyond it.

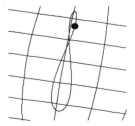

Figure 2.5 The Sun's analemma

THE ECLIPTIC

Just as with the motion of stars, we have first discussed the *observed* motions of the Sun. What comes next is to try to construct a model that explains *why* the Sun appears to move the way it does.

Since the Celestial Sphere model worked so well in explaining the motion of the stars, it can also be used for the motions of the Sun. Stars do not move relative to one another, they each have a fixed point on the celestial sphere. When observed from a given location, a given star will follow the same path every day. Its rising and setting positions and transit altitude do not change. However, as we have seen, the Sun's path is slightly different each day, the full cycle of changing paths defining a

15

year. This cycle requires the Sun to have a *path* on the celestial sphere rather than just a point like each star.

A path extending all the way around the celestial sphere tilted at an angle of 23 1/2° to the celestial equator and intersecting it at two points will result in motions of the Sun that agree with our local observations. This line is called the **ecliptic**, the Sun's path through the stars.

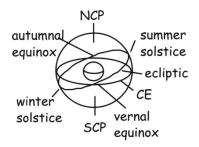

Figure 2.6 The celestial sphere and the ecliptic

On any given day, as the celestial sphere appears to rotate, it "takes" the Sun, in its current position, with it. This results in the Sun's daily motion across the sky. By the next day, the Sun will have moved a small amount along the ecliptic and the path will be a little bit different. As the Sun moves around the ecliptic once a year, it will be observed to go through its entire cycle of changing rising and setting positions and transit altitudes.

The northernmost point on the ecliptic is the summer solstice. The southernmost point is the winter solstice. The points in between, the intersection points between the ecliptic and the celestial equator, are the equinoxes. These points are named for the seasons in the Northern Hemisphere. When the Sun is at its northernmost point, it is high in the Northern Hemisphere, so it is summer there, but it is low in the Southern Hemisphere, making it winter there. The reverse is true on the winter solstice. All seasons in the Southern Hemisphere are opposite those in the Northern Hemisphere.

16

The direction the Sun appears to move along the ecliptic is opposite the apparent rotation of the celestial sphere, so each day the Sun will appear to fall a little bit behind the stars in its daily path by about 4 minutes. Because we base our daily routines on the Sun and not the stars, our time measurements are based on the solar day. Since we have defined a solar day as 24 hours, the sidereal day will be about 23 hours and 56 minutes. This means that a star will appear to rise 4 minutes earlier each day. In one day this will not be very noticeable, but in a week the star will rise about 1/2-hour earlier, in a month, 2 hours earlier, and in a season (3 months), about 6 hours earlier. Eventually it will be up with the Sun and you will no longer be able to see it because the Sun is so bright that stars that are up at the same time are lost in the glare. They are still there, but you cannot see them.

Stars visible at certain times of year will not be visible at other times. When the Sun is in a certain position on the ecliptic, stars and constellations on the celestial sphere near that location will not be seen because they are up *with* the Sun. The stars and constellations you *will* see are those on the opposite side of the celestial sphere from the Sun. As the Sun moves around the ecliptic throughout the year, different stars become visible. This is the reason for constellations being seasonal. We consider a constellation as part of the seasonal sky in which we can see it best. Seasonal star charts are provided in **Appendix 2**.

THE ZODIAC

The stars that lie on or around the ecliptic are known as those of the **zodiac**. The zodiac is divided into 12 constellations. The Sun spends about 1 month a year "in" each constellation. Watching the Sun traverse the ecliptic and move through the zodiac is yet another way to experience the cycle of a year.

SEEING THE SUN

As with the stars, the celestial sphere can be used to determine where and when an observer will see the Sun from various places on the Earth at various times of the year. Remember from **Chapter 1** that by looking overhead an observer establishes his or her own zenith. Their latitude will also be the declination directly overhead, and everything within 90° of this declination will be visible above their horizon.

Observers at the poles always see the same half of the celestial sphere, so they will only see the Sun when it is in the half visible to them. At the North Pole the Sun is up all the time during the spring and summer and down during the fall and winter. The exact opposite is true at the South Pole.

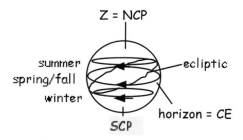

Figure 2.7 The Sun's paths seen from the North Pole

Any location within 23 1/2° of the poles will have at least one day of all daylight in its summer and one of all darkness in its winter. As you move toward the pole, this time period increases to a few days, weeks, months, and finally 6 months, exactly half the year at the pole. The lines of 66 1/2° north and south latitude are called the **Arctic** and **Antarctic Circles**. All locations near these latitudes have extreme seasonal variations in amounts of daylight and darkness. However, as can be seen in Figures 2.7 and 2.8, the altitude of the Sun is never very high. As a result, it is generally cold at these latitudes.

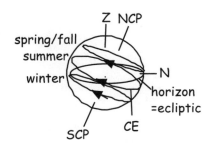

Figure 2.8 The Sun's paths seen from near the Arctic Circle

In terms of sunlight duration and angles, the tropical and equatorial regions of the Earth are opposite the polar and arctic regions. The lines of 23 1/2° north and south latitude are known as the **tropics** of **Cancer** and **Capricorn**. These are the latitudes that the Sun is directly above on the days of the solstices. Therefore, on June 21, observers along the tropic of Cancer will see the Sun directly overhead at local noon, and the same is true for the tropic of Capricorn on December 21. The Sun crosses the celestial equator on the equinoxes, so when observed from the Equator at noon on March 21 and September 21 the Sun will reach the zenith.

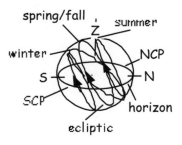

Figure 2.9 The Sun's paths seen from the Tropics

As seen in Figure 2.9, even in the winter the Sun is as high in the Tropics as it is during the spring in the middle latitudes. Since the Sun's altitude is always high in the equatorial regions, the weather is generally warm there. However, just as with the stars, the Sun is both up and down for 12 hours each every day of the year (see Figure 2.10). In the Tropics there is a small variation in the duration of sunlight that is less than in

19

the middle latitudes. The seasonal variations in daylight and darkness are the least at the equator and increase as you move toward the poles.

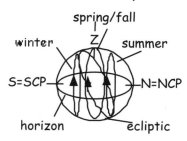

Figure 2.10 The Sun's paths seen from the Equator

THE SUN AND EQUATORIAL COORDINATES

The equatorial coordinate system discussed in **Chapter 1** is partially defined by the ecliptic. The celestial equator is a natural choice for the 0° line of declination, just as the equator is the obvious line of 0° latitude in geographic coordinates. However, the choice of 0 hours right ascension is as arbitrary as the 0° line of longitude, the prime meridian. The line of right ascension that goes through the vernal equinox is designated as 0 hours right ascension. This designation therefore makes the line of right ascension going through the summer solstice 6 hours RA and the lines through the autumnal equinox 12 hours RA and the winter solstice 18 hours RA.

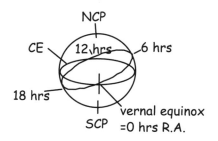

Figure 2.11 The equinoxes and solstices on the celestial sphere

Flat Maps

Maps of the Earth in geographic coordinates are often flattened-out. This is just a matter of convenience (a globe wouldn't fit very well in a glove compartment during a vacation). On these flat maps, lines of latitude run across the page and lines of longitude run up and down. The equator divides the map into a northern and a southern half, and the map must be "cut" along some line of longitude, often the International Date Line, in order to be flattened-out. It is important to remember that locations on the extreme east and west sides of the map are really right next to each other.

This flattening is also done with maps of the sky. Lines of declination run across the map, and lines of right ascension run up and down. Usually the "cut" is made at the right ascension of the autumnal equinox, so the center of the map is the right ascension of the vernal equinox. Just as with geographic maps, the celestial equator divides the map into a northern and a southern half. The ecliptic's tilt relative to the celestial equator causes the Sun's path through the sky to appear as a curved line above the celestial equator on the east (left) side of the map and below it on the west (right) side of the map. To understand why east and west are opposite the directions on a map of the Earth, think of laying on the ground looking at the sky with your head to the north and feet to the south. East would be to your left and west to your right.

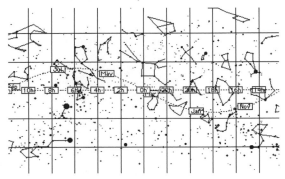

Figure 2.12 A flat star map

21

Activities

1. Visit a planetarium to observe the motions of the Sun discussed in this chapter.
2. Better yet, go outside and see the motions yourself. Observational Activities can be found in **Appendix 1**.
3. Complete **Celestial Globe Activity #2-The Sun and the Seasons**, found in **Appendix 3**.
4. Complete **Celestial Globe Activity #3-Around the World**, found in **Appendix 3**.

Exercises

THE SKY WHEEL

The Sky Wheel can be used as an aid in understanding the observed motion of stars and the Sun from different locations on the Earth throughout the year. Trace "Local Horizon" and "Sun's Path" on the next page onto transparency (do each in a different color if you can). Cut the transparency so the "Local Horizon" and the "Sun's Path" can be placed over the "Celestial Sphere" separately.

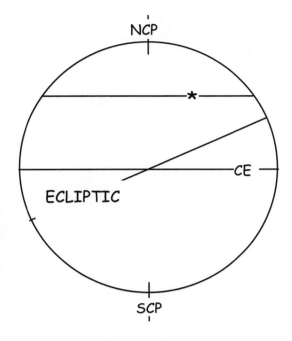

NCP

*

CE

ECLIPTIC

SCP

Celestial Sphere

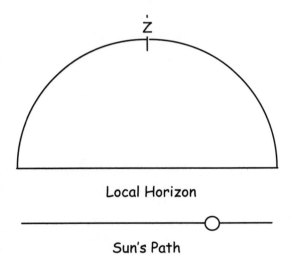

Z

Local Horizon

Sun's Path

MOTION OF THE STARS

1. Point the zenith of the "Local Horizon" to whatever declination on the "Celestial Sphere" is right above the latitude from which you wish to observe. Latitude is always equal to the declination overhead.
2. Rotate the "Local Horizon" and "Celestial Sphere" together until the zenith is pointing directly upward.
3. Whatever is within the lines of the "Local Horizon" is the part of the "Celestial Sphere" that can be seen from the latitude you chose.
4. You can now observe the altitudes of the NCP and CE and the path of the star from any location.

MOTION OF THE SUN

1. Place the "Sun's Path" anywhere on the "Celestial Sphere" as long as it is parallel to the CE and touching the ecliptic. The "Sun's Path" touching the northernmost point on the ecliptic (the point closest to NCP) represents the summer solstice; the southernmost is the winter solstice.
2. Now follow steps 1 and 2 in Motion of the Stars to observe the Sun's path for any given location at any time of the year. Observe whether the path is closer to the zenith or the horizon; observe whether the Sun spends more time above or below the horizon, that is, whether more of the Sun's path is above or below the horizon.

Questions

Use the Sky Wheel to answer the following questions:

1. At the North Pole, what part of the celestial sphere is observed at the zenith? What is on the horizon? How are the paths of the stars oriented to the horizon?
2. Answer the same questions as in Exercise 1 for the equator.
3. Where are the NCP and the CE observed from your latitude? How are the paths of the stars oriented to the horizon?

4. Observe the path of the Sun from your latitude on both the summer and winter solstices. What differences do you see in the Sun's path? There should be three obvious ones.

5. Compare the Sun's path at your latitude to paths at latitudes both north and south of you at a given time of year. What differences do you see?

6. Place your zenith at a location just south of the North Pole. Observe the Sun's path at both the summer and winter solstices. What do you notice that you do not see from a midlatitude?

7. What time(s) of year is it when the Sun's path is on the point of the ecliptic that intersects the CE?

8. Place your zenith at a Southern Hemisphere latitude similar to yours. When is the Sun's path highest? When is the Sun's path longest? In what direction would you have to look at noon to see the Sun? How does this compare to your Northern Hemisphere location?

9. From what locations on Earth could you observe "Midnight Sun?" When would this occur?

10. Can you observe the Sun at the zenith from your location? If not, when is it closest to the zenith? Where do you have to go to see the Sun on the zenith?

11. If there were no Sun at all in a day at a location somewhere above the Arctic Circle, what would the day be like the same distance below the Antarctic Circle?

12. If it were the "longest day" of the year in Detroit, Michigan, 42°N latitude, what would the day be like in Wellington, New Zealand, 42°S latitude?

Chapter Three

POSSIBLE EXPLANATIONS

Thus far, the observed motions of the stars and the Sun as seen from various locations on the Earth at different times of the year have been discussed. The celestial sphere was introduced as a possible explanation for these motions.

The celestial sphere is an example of a **scientific model**. Whether or not a scientific model represents what is actually physically occurring is not necessarily the first concern; a model can be considered successful if it works. The celestial sphere works because it can be used to explain the observed motions of the stars and the Sun from various locations on Earth.

In the celestial sphere model the Earth is stationary and all objects appear to move around it. The sky appears to rotate around the Earth once per day. The Sun seems to move with the sky each day, but also appears to revolve around the Earth on the ecliptic once per year. This model is Earth-centered or **geocentric**.

Although the geocentric explanations work, there are alternate explanations. These are Sun-centered, or **heliocentric**, explanations. Helios was the character in Greek Mythology that rode the Sun's chariot across the sky each day.

In the heliocentric model, the daily motion of the sky is attributed to the Earth rotating once per day "underneath" a stationary sky. The sky therefore appears to be moving opposite the direction of the Earth's rotation. Since the stars and the Sun rise in the East and set in the West, the rotation of the Earth must be from west to east. This is just like riding a merry-go-round. Everything seems to be moving around you in the direction opposite your motion.

The Sun's annual motion is attributed to the Earth revolving around the Sun once per year, making the sun appear to be moving through the stars.

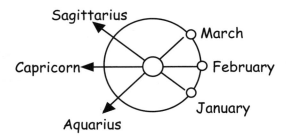

Figure 3.1 Earth's orbit makes the Sun appear to move through the stars

The seasonal changes in the Sun's daily path are explained in the heliocentric model by attributing the tilt to the Earth's rotational axis. As the Earth revolves around the sun once per year, the tilt stays in the same direction, toward Polaris, and different parts of the Earth receive different amounts of sunlight.

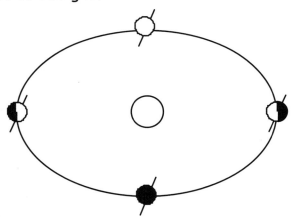

Figure 3.2 The Earth's tilted rotational axis causes the seasons as the Earth orbits the Sun (sizes and distances in figure not to scale)

Note the Sun angles at different locations in Figure 3.3, which represents Northern Hemisphere summer. The Sun's rays appear high (making large angles with the Earth's surface) in the equatorial and tropical locations, while low (making small angles with the Earth's surface) near the Poles. The rays are also high in the northern midlatitudes, while low at similar southern latitudes. Also note that as the Earth rotates, the northern locations will spend more time in daylight than the southern ones, including the North Pole, which is in sunlight all day, while the South Pole is in darkness. The opposite is true in the winter.

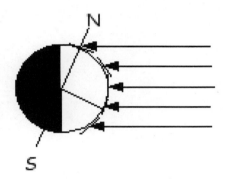

Figure 3.3 A close up of Northern Hemisphere Summer

Questions

1. Can you tell by observation alone which model is "right?" That is, which model is more likely to represent physical reality or what is really happening? If so, explain the observations.

2. There is a principle in science known as "Occam's Razor." Basically, it states that when all else is equal, the simpler explanation tends to be the correct one. Employing Occam's Razor, which model is simpler and therefore preferable?

Comet LINEAR. Taken at Edinboro University of Pennsylvania-Maize
Sunfire Observatory. Used with Permission of the Director.

Chapter Four

MOTIONS
OF THE MOON

The Moon is the object in the sky that goes through the fastest changes. These changes are in both appearance and position. We see varying amounts of the Moon's surface from day to day. These changes are known as different **phases**. The times that the Moon rises and sets also change. They are almost an hour earlier each day and are coordinated with the phases.

LUNAR PHASES

The Moon cycles through its phases and variations in rising and setting times in a period of about 29 1/2 days. This time period is called a **synodic month**. The "mon" in month stands for Moon ("month is easier to say than "moonth"). The approximate 30-day length of our calendar's months comes from this length of the cycle of the Moon's phases.

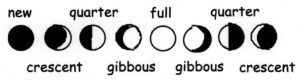

new   quarter   full   quarter

crescent   gibbous   gibbous   crescent

Figure 4.1 Lunar phases

From the new to full phases, the Moon is getting brighter, so these phases are referred to as "waxing." The phases from full to new, when the Moon is getting dimmer, are called "waning" phases.

Whether the perspective is geocentric or heliocentric, the observed motion of the Moon can be explained by it being in orbit around the Earth. As the Moon orbits the Earth, the sides of both objects facing the Sun will be illuminated; the sides opposite the Sun will be dark. To determine how the Moon will look from Earth when it is in any given position in Figure 4.2, look at Figure 4.3. Imagine yourself on the Earth below any position of the Moon. Then draw a line from your position on the Earth to the center of the Moon. Draw another line, perpendicular to the first, also through the center of the Moon. This line will show you which part of the Moon's surface will appear illuminated to an observer "under the Moon" on Earth. The procedure shown in Figure 4.3 can be followed for any of the eight positions of the Moon in Figure 4.2 and should result in the phases pictured in Figure 4.1. You will use this procedure to complete an exercise at the end of this chapter.

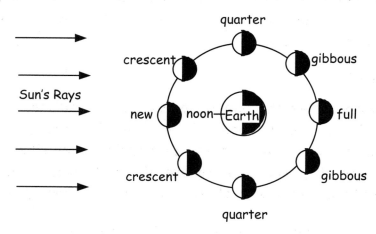

Figure 4.2 The Moon's orbit viewed from above

The times that the Moon will be visible can also be determined from Figure 4.2. At the point on the Earth directly below the Sun, it is 12 noon. At the point opposite noon, it is 12 midnight. Viewed from above the North Pole, the Earth rotates counter-clockwise. This defines the time at all other points on the Earth. For instance, at the point half way between noon and midnight, it is 6 PM, and at the point half way between midnight and noon, it is 6 AM. The time on this "Earth-clock" that the Moon is directly above is the time that the Moon will be seen is at its highest altitude in the southern sky. For simplicity, the rising time of the Moon can be considered six hours before it is "overhead" and the setting time six hours after it is "overhead" (12 hours after it rose). These rising and setting times can be determined for any position of the Moon in Figure 4.2. You will also do this in an exercise at the end of this chapter.

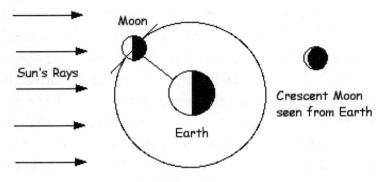

Figure 4.3 Determining the phase of the Moon

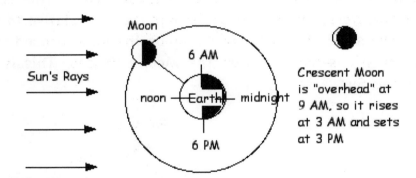

Figure 4.4 Determining the rising and setting times for the Moon

Note that inspection of Figure 4.2 shows that when the illuminated side of the Moon is facing the Earth (during the bright phases like full and gibbous) the Moon is mostly visible at night. When the Moon is on the other side (in the dimmer phases like crescent) it is mostly seen during the day.

MONTHS

Although the synodic month (the cycle of the Moon's phases) is caused by the Moon revolving around the Earth, 29 1/2 days is actually a little bit more time than the Moon requires to orbit the Earth. An observer at position A in Figure 4.5 will observe a full Moon in a certain position relative to the stars. As the Moon orbits the Earth, the Earth is also orbiting the Sun. After one complete revolution around the Earth, the Moon returns to the same position in the sky relative to the same star, but only about 27 1/3 days have past. This is because the Earth has moved about 1/12 of its way in orbit around the Sun (there are 12 months in a year). An observer at point B sees the Moon in the same location relative to the stars as from point A, but will *not* see a full Moon. The observer at point B will see a gibbous Moon. It will take another 2+ days for the Moon to reach full again. The 27 1/3 days that the Moon takes to revolve around the Earth is called a **sidereal month**. This is a month relative to the stars. The Moon's position in the sky relative to the stars will appear the same after one revolution of the Moon around the Earth. A synodic month is the cycle of the Moon's phases. This is a month relative to the Sun.

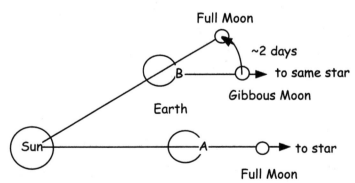

Figure 4.5 The sidereal and synodic months

34

The heliocentric explanation for the difference between the sidereal and solar days is similar to that of the two types of months. It is left as an exercise at the end of this chapter for you to diagram the motions of the Earth and try to explain the reason for the difference.

THE FAR SIDE OF THE MOON

As the Moon orbits the Earth, the same features are always seen on its disk. In other words, the Moon always faces the same side towards the Earth. There is a "far side" that an observer from Earth can never see. Apollo spacecraft photographed it in the late 1960s, but it is impossible to see the far side from Earth.

The reason for this is that the Moon rotates in exactly the same amount of time, 27 1/3 days that it takes to revolve around the Earth. This **synchronous rotation**, or 1:1 resonance as it is often called, is shown in Figure 4.6 and can be verified by an activity at the end of the chapter.

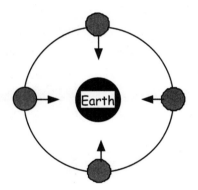

Figure 4.6 The same side of the Moon always faces the Earth

ECLIPSES

From our point of view the Sun and the Moon both appear to be about the same size. Which is actually larger depends on which one is farther away. In angular size, they appear the same to us on Earth, about 1/2°. An entire circle is 360°, so 720 disks of the Sun or the Moon could fit around the whole sky. Opposite points on the horizon are 180° apart. If you are facing one way it is 180° to the point behind you. From the horizon to the zenith is 90°. Your fist held out at arm's length is about 10°, your thumb is about 2°, a finger 1°, and your pinky about 1/2°. This means that your pinky held out at arm's length could block out the Sun or the Moon. Try it. This is true whether the Sun and the Moon are high overhead or close to the horizon. The Sun and the Moon often appear much larger when close to the horizon since there are nearby objects on the horizon that they can be compared with, but they are both 1/2° in size no matter where they are. This can be tested. You will find that no matter how big either object looks; they can still be covered by your pinky. Try it!

The coincidence that the Sun and the Moon have the same angular size is the reason that events known as eclipses are so interesting. As indicated by the name, an **eclipse** can occur when the Moon is on or near the ecliptic. The ecliptic, in geocentric terms, is the Sun's apparent path through the sky. In heliocentric terms, it can be thought of as the line between the Sun and the Earth. If the Moon is on the ecliptic when it is in between the Sun and the Earth (new Moon), it will block out the view of the Sun from the Earth. This is a **solar eclipse**. If the Earth is between the Sun and the Moon (full Moon), the Earth's shadow will cover the Moon. This is a **lunar eclipse**.

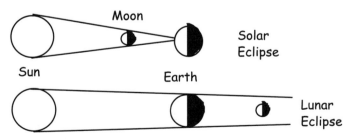

Figure 4.7 Solar and lunar eclipses

Eclipses do not occur at every new and full Moon. This is because the Moon's orbit is tilted about 5° from the ecliptic. The ecliptic and the Moon's orbit are shown in the geocentric perspective in Figure 4.8. The intersection points are called **nodes**. The line passing through the Earth and connecting the nodes is known as the **line of nodes**. When the Sun and the Moon are both on the line of nodes at the same time, an eclipse will occur. The Sun traverses the ecliptic once per year, so it will be on the line of nodes twice per year. The Moon orbits the Earth once per month, so it will be on the line of nodes 24 times a year. If the Moon crosses a node when the Sun is on one there will be an eclipse. If the Sun is not on a node there will not be an eclipse. Since the Moon travels through the sky about 12 times as fast as the Sun, it usually comes near enough a node whenever the Sun is on one so that eclipses generally occur about twice per year. The times that the Sun is on or near a node are called "eclipse-seasons." If the Sun and Moon are at the same node, a solar eclipse occurs. If they are at opposite nodes, a lunar eclipse occurs. If the Sun and the Moon are not exactly lined up, but are close, a partial eclipse will occur.

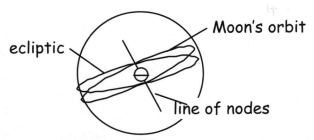

Figure 4.8 Eclipses occur when both the Moon and the Sun are on the line of nodes

Referring back to Figure 4.7 shows that most people are likely to see more lunar eclipses in their lifetime than solar eclipses. During a lunar eclipse anyone on the entire night side of the Earth (the side facing away from the Sun) can see the full Moon eclipsed. During a solar eclipse only people in the specific position on the Earth covered by the Moon's shadow will see the eclipse.

Activities

1. Go outside and observe the Moon's phases for real. **Observational Activities** are included in **Appendix 1**.
2. Complete **Celestial Globe Activity #4-The Moon**, found in **Appendix 3**.
3. *Do-it-Yourself Phases*-Use a ball for the Moon, a lamp for the Sun, and yourself for the Earth. Turn on the lamp and slowly move the ball around you and observe the ball going through the same phases as the Moon. Also watch for eclipses!
4. *Do-it-Yourself "far-side" Demonstration*-Pretend that one student in the class is the Earth and you are the Moon. Walk in orbit around the other student while always keeping your face towards them. This is the same thing that the Moon does when it orbits the Earth. Are you rotating? If you are, what is similar about your rotation compared to your revolution?

The surface of the Moon. Taken at Edinboro University of Pennsylvania-Maize Sunfire Observatory. Used with Permission of the Director.

2.   Construct a diagram similar to Figure 4.3 in this chapter that explains
       the reason for the 4-minute difference between the solar and sidereal
       day.

Questions

1. Explain how using the term "dark side" of the Moon could cause confusion when referring to the Moon's far side.

2. Refer to Figure 4.8. What kind of eclipse would be observed from the Moon while the Earth is experiencing a lunar eclipse? What would be observed from the Moon while the Earth is experiencing a solar eclipse?

3. Does a New Moon occur on the same day of every month on the calendar? Explain your answer.

4. Have you ever seen the Moon during the day?

Chapter Five

PRECESSION, TIDES AND THE CALENDAR

PRECESSION

The two motions discussed thus far are **rotation**, an object spinning on an axis, and **revolution**, one object in orbit around another. Rotation is responsible for the daily motion of the sky. Whether the rotation is attributed to the celestial sphere or the Earth, the observed motion, the stars and the Sun rising in the East and setting in the West, is the same. Revolution is responsible for annual motion in the sky. Once again, whether the revolution is considered the Sun around the Earth on the ecliptic or the Earth around the Sun, the observed changes in the daily path of the Sun and the stars visible at night will be the same.

A third motion that occurs is called **precession**. Precession is probably most familiar as the motion of a top when it is running out of energy and about to fall. The rotational axis traces out the path of a cone. The North Celestial Pole, the point directly above the Earth's North Pole, traces out a circular path through the stars about once every 26,000 years. As with the other motions, the precession could be attributed to either the celestial sphere or the Earth. Although it takes 26,000 years, much longer than rotation or revolution, the effects of precession can be noticed in much smaller amounts of time. Since precession actually changes the position of the North Celestial Pole, the star located at or closest to this point will not always be the same. Polaris is our current "Pole Star" and obviously has been ever since it was first named for its position. However, other stars have been the Pole Star in the past and others will be in the future. Figure 5.1 shows the path of the precession of the pole.

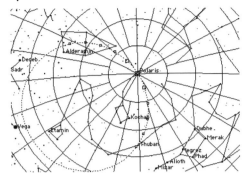

Figure 5.1 The dotted line is the path of the precession of the pole

Changing the direction that the Earth's axis is pointing also affects our perception of the seasons. Our calendar is based on the observed motion of the Sun. If precession occurred without adjustment of the calendar, the seasons would appear to be occurring in months different than we expect them to.

Figure 5.2 Precession

This is most easily visualized from a heliocentric perspective. For instance, when the calendar reads July, the Northern Hemisphere is pointed toward the Sun and experiences high Sun angles and long daylight periods resulting in warm summer weather (as discussed in **Chapter 3**). After about 13,000 years, half of a complete precession, the Northern Hemisphere would be pointing away from the Sun at that particular position in its orbit. Summer would still occur, but when the Earth is on the opposite side of the Sun. So, according to the calendar, summer would occur in January and winter in July. This effect is noticeable in much less than 13,000 years, so minor adjustments are periodically made in the calendar to compensate for precession and keep the months in the seasons we are used to.

CALENDARS

Cycles in the sky have been the basis of calendars since prehistoric times. Many archeological sites, such as **Stonehenge**, on the Salisbury Plain in England (see cover photo), have been theorized to be calendars of some sort. Since the daily routine of most people is based on the motion of the Sun (sleeping at night and being active during the day) it is logical that the solar day is the unit of time on which most calendars are based. Similarly, the tropical year is a logical cycle to use as a longer unit of time. As mentioned in **Chapter 2**, there are about 365 1/4 solar days in a tropical year, which requires every fourth year to be a "leap year" with an extra day.

The synodic month, the 29 1/2 days that the Moon takes to go through its phases, discussed in **Chapter 4**, is also a useful larger time unit. Some calendars are lunar and not solar, that is, based on the cycles of the Moon rather than the Sun. Muslim and Jewish religious calendars are examples of this. For instance, the Muslim Holy Month of *Ramadan* appears to "drift" through the solar calendar. It begins 11 days earlier in the solar calendar each year because 12 synodic months (12 x 29 1/2= 354 days) is 11 days less than a solar tropical year (12 days in leap years). Even when rounded off to 30 days, a whole number of synodic months will not fit into the 365 or 366 day year. So the approximate length of the months is the only lunar influence on solar calendars.

The first calendar used by most of western civilization was the **Julian calendar**, named after Julius Caesar since it was adopted during his reign (around 45 BC). Earlier Roman calendars had only 10 months. The months September, October, November, and December were originally the seventh through tenth months of the year. The year began in the spring with March, followed by April, May, and the fifth and sixth months then called Quintilis and Sextilis. Our current winter months of January and February were originally a period of waiting for spring and were added later. The months not named for their order within the year were named after mythological figures.

Each month in the Julian calendar was 30 days. The year began after a 5 day holiday (6 in leap years) after the winter solstice. This holiday was called "*Kalends*," which is the origin of the word calendar. It was also the beginning of many of the traditions that can now be found as part of year-ending holidays such as *Christmas* and *Chanukah*. Julius Caesar decided to honor himself by changing the name of Quintilis, which was then the seventh month, to July and stealing a day from February to make his month the longest. Not to be outdone, his successor, Augustus, did the same thing with Sextilis, which is now our 31-day August.

Unfortunately for the Julian calendar, the year is not exactly 365 1/4 days. It is close, but it is actually 365.2422 days and not 365.2500. This 0.0078 day difference (about 11 minutes and 14 seconds) may not seem like much, but over many years it adds up. By the Middle Ages, farmers started noticing spring, the time to plant crops, coming earlier than they expected from reading the calendar. This was also true of the first frost in the fall, before which many of the crops had to be harvested. Also, the times of the solstices and equinoxes were occurring earlier than their expected calendar dates. This discrepancy necessitated a change in the old Julian calendar. In 1582, during the reign of Pope Gregory, the calendar was advanced by 11 days, the amount that the Julian calendar had fallen behind in the roughly 1,500 years since it had been established. Also, it was decided that years not divisible by 400 would NOT be leap years. For instance, 2000 was a leap year, but 1700, 1800, and 1900 were not. Skipping 3 leap years every 400 years will keep the new **Gregorian calendar**, as it is now called, accurate for over 3,000 years.

An amusing story goes with this change. Apparently people were protesting at the Vatican that when "fixing" the calendar, Pope Gregory had taken away 11 days of their life. Some argued that they should be compensated by a lump-sum payment of 11 day's wages. Problems also occurred when tenants and landlords could not agree on whether a whole month's rent should be paid for the October in which this occurred. Whether banks should charge or pay a full month's interest was an issue as well. Also, since the Pope's authority was not (and is still not) worldwide, some countries did not accept the Gregorian calendar right away; in fact, some countries did not adopt it until the early 1900s.

THE TIDES AND THE DAYS OF THE WEEK

The length of the day and the year are based on motions of the Sun. The length of the month is based on motions of the Moon. Did you ever wonder why a week is seven days?

It is possible that the week is not astronomically based at all, but rather that it comes from the creation story of Genesis in the Judeo-Christian Old Testament. It states that God created the world in 6 days and took the seventh to rest. One astronomical perspective is that it takes about 7 days for the Moon to cycle through a quarter of its phases. This is also the amount of time that ocean tides (which are caused by the Moon and the Sun) take to cycle from their greatest to their smallest difference between high and low tide. The greatest difference between high and low tide is called "spring" tide and occurs during the full and new Moon. The smallest difference is called "neap" tide and occurs during the quarter phases.

Tides are caused by the difference in gravitational pull from the Moon and the Sun from one side of the Earth to the other. This is also the cause of Precession. Gravity will be discussed in more detail in **Chapter 10**. The side of the Earth facing the Moon or Sun will be pulled with greater force than the opposite side. This differential gravitation causes a "tidal-bulge." Since water is a very fluid substance, the Earth's oceans will be deeper along the line of the tidal bulge, high tide. The excess water for high tide comes from the surface locations perpendicular to the bulge, which is therefore at low tide. Even though the Sun is much more massive than the Moon, the Moon is much closer, so the differential in the gravitational pull from one side of the Earth to the other due to the Moon is more than that due to the Sun. Therefore, the location of the Moon determines the locations of high and low tides and the location of the Sun will determine whether it is spring or neap tide.

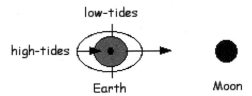

Figure 5.3 A view from above the North Pole. High tides are found on the Earth's surface along the line to the Moon, low tides along the line perpendicular to the line to the Moon.

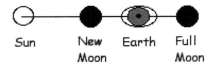

Figure 5.4 Spring tides occur during Full and New Moons. Notice that the Moon and the Sun are both causing tides along the same line; they are "working together." At these times there are higher high tides and lower low tides resulting the largest difference between high and low tide.

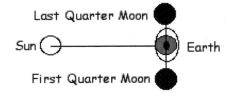

Figure 5.5 Neap tides occur during the Moon's Quarter phases

Notice that the Moon and the Sun are causing tides along perpendicular lines. The high tides will be along the line from the Earth to the Moon, due to its stronger differential gravitational pull, but lower

48

due to the Sun "working against" the Moon pulling along the line from the Earth to the Sun, where the low tides will be. At these times there are lower high tides and higher low tides giving the smallest difference between high and low tide.

Another possible explanation for the 7-day week is that one day was named for each of the 7 objects visible to ancient astronomers (without telescopes) that moved relative to the fixed background of stars. You know them as the Sun, the Moon, and the five planets, Mercury, Venus, Mars, Jupiter, and Saturn. The English names for some of the days reveal this, such as Saturday, Sunday, and Monday. Latin languages such as French and Spanish show even more direct correlation, see Figure 5.6.

Some of the English and German names of the days correspond roughly to Norse equivalents of the Roman mythological characters after which the planets were named. Tuesday is named for Tiw, the Norse god of War, like the Roman Mars. Wednesday is named for the Norse Woden, who dispensed justice, like the Roman Mercury. Thursday is the thunder god Thor's day. Donner is the German word for thunder, so Donnerstag literally translate to "thunder's day." The Roman god who made thunder was Jupiter. Friday comes from Fria, the Norse equivalent of Venus, the Roman goddess of love and beauty.

Object	English	German	French	Spanish
Sun	Sunday	Sonntag	Dimanche	Domingo
Moon	Monday	Montag	Lundi	Lunes
Mars	Tuesday	Dienstag	Mardi	Martes
Mercury	Wednesday	Mittwoch	Mercredi	Miercoles
Jupiter	Thursday	Donnerstag	Jeudi	Jueves
Venus	Friday	Freitag	Vendredi	Viernes
Saturn	Saturday	Samstag	Samedi	Sabado

Figure 5.6 The names of the days of the week in different languages and the objects they are named for.

Exercises

1. The path of the precession of the pole in Figure 5.1 represents 26,000 years of precession, and the direction of the precession is clockwise. Determine which star was the Pole Star in 3000 BC. The Great Pyramids of Egypt were built then, approximately 5,000 years ago, and are pointed to this star. As you learned in **Chapter 1**, Duhbe and Merak are used as "pointer stars" to our current Pole Star, Polaris. Use Figure 5.1 to determine which stars would have been the pointer stars to the pole in 3000 BC.

Questions

1. Explain the affect that precession can have on a calendar.
2. Since the Earth rotates once in about 24 hours, how much time is there between high and low tide at a given location? How many high tides and how many low tides will a location experience in a day?
3. Name the motion responsible for each of the following time measurements: day, week, month, and year.

Chapter Six

THE GEOCENTRIC SYSTEM

PREHISTORIC ASTRONOMY

Humans have shown an interest in astronomy since prehistoric times. Stonehenge, mentioned in the last chapter, is only the most famous of many prehistoric relics that show an astronomical consciousness. Ruins all over the world, including in the United States, show that our ancestors were watching the sky.

This arguably makes astronomy the oldest of the sciences. Why were our ancestors such avid sky watchers? Understanding the sky's cycles was literally a matter of life and death. As discussed in the last chapter, our calendars and clocks are based on cycles in the sky. Why did ancient people care about measuring time? Many of our most distant ancestors lived in nomadic tribes. They were nomads because they moved with food supplies, which probably also meant moving with warmer weather as well. Certainly weather was as unpredictable in the past as it is now, so if people waited for weather changes, they may have moved too late. Watching the sky was much more reliable. The annual reappearance of a certain bright star, a specific position of the Sun or the Moon could have indicated that it was time to move on. It is also possible that they used the positions of certain stars, such as the Pole Star or others, to navigate.

The advent of agriculture allowed people to stop wandering and stay in one place. This was the beginning of civilization. Watching the sky continued to be just as important so people knew when to plant and harvest crops and when to seek warmer clothing and shelter.

Many ancient civilizations were geographically isolated from one another, and communication, other than through wars and conquest, was minimal. As a result of this isolation there was very little exchange of knowledge and discoveries of any type. This, of course, included science. It was not until only about 2,500 years ago that enough information exchange occurred to establish "worldviews" in various disciplines. A worldview is an idea or belief about something that is shared by a significant number of peoples in different places. The first worldviews of science and specifically astronomy came from the ancient Greeks.

GREEK ASTRONOMY

In ancient Greece, intellectuals were called philosophers. There were different kinds of philosophy; the study of the world and its workings was called natural philosophy. The word "science" comes from the Greek for "natural philosophy." Considered one of the greatest of all the natural philosophers, **Aristotle** is often revered as one of the world's first scientists.

Aristotle had many ideas about the Earth. First he thought that the Earth did not move. He felt that if it did we should feel it; there would always be a wind opposite the direction of the motion, like in a ship on the Aegean Sea. He also believed that the Earth was round. The introduction of the idea of a spherically-shaped Earth is usually attributed to Columbus during the Renaissance. However, that idea, along with many others, were things that were known during the time of classical civilization and had to be rediscovered after the Dark Ages.

Aristotle also thought that the Earth was central to the cosmos (or universe, as we would say today). Why did he think that? Based on the discussions of the observed motions of the stars, the Sun, and the Moon in previous chapters, it should be obvious that thinking the Earth is central is a very natural idea.

For the next 500 years, different Greek astronomers studied the heavens. There were many contributors to this first worldview, but we owe much to the one who recorded it so it could be passed down to subsequent generations, Claudius Ptolemy. Ptolemy lived in Alexandria, Egypt during the 2nd century AD. This was not Egypt of the Pharaohs, but rather Roman-dominated Egypt. Around 150 AD Ptolemy compiled much of what was known about ancient Greek astronomy in a work called *The Almagest* (A combination Greek-Arabic word for "The Greatest"). This may sound a bit conceited, but as we will see, *The Almagest* contained what were the greatest theories about astronomy, and perhaps even all of science, that the world had yet seen.

Figure 6.1 Aristotle and Ptolemy

THE GEOCENTRIC SYSTEM

Ancient Greek astronomy basically consisted of the geocentric explanations for the motions of the stars and the Sun discussed in earlier chapters. Many, including Ptolemy, just considered the ideas devices for explaining observations and were not concerned with whether the explanations represented physical reality. Either way, the development of the explanations in the geocentric model or system may have gone something like what follows.

The first observations showed that the stars seemed to be all moving together in circular paths around the Earth. So a logical explanation was that they were all fixed to a gigantic crystal sphere that was centered on the fixed Earth and rotated once per day.

Further observations designed to test the idea of the crystal sphere showed that the Sun and the Moon moved independently of the stars and therefore needed their own spheres. This makes the model a bit more complex, but a model is useless unless it explains all observations. Both spheres must have been in front of the background stars they appeared to move through, the Moon sphere being closer to the Earth. There were two good reasons for this. Thinking about what they were is left as an exercise at the end of this chapter.

Next, when observations were made to test ideas of the Sun's and the Moon's spheres, it was noticed that there were several objects in the sky that looked like stars (points of light, although brighter than most) that moved with their own paths, more like the Sun and the Moon. These 5 objects were referred to as the "wanderers." *Planet* is the Greek word for wanderer. These planets were problematic for Greek astronomers because they were making the model even more complex. Simplicity is always desirable in a scientific model (see Occam's Razor in **Chapter** 3). Whenever they studied them, the astronomers said they were working on the "problem of the planets." Five spheres had to be added to the model, one for each planet. The planets were named for the Roman gods Mercury, Venus, Mars, Jupiter, and Saturn.

The first problem would be what order to put them in; determining how the order listed above was decided upon is also left as an exercise at the end of this chapter. One important distinction was that observations showed that there were clearly two types of planet. Mercury and Venus were never seen very far from the Sun in the sky, only in the early morning right before sunrise or the early evening right after sunset. Because of this, they were considered **inferior** or "below the Sun," closer to the Earth than the Sun. Mars, Jupiter, and Saturn, however could be seen at anytime, sometimes close to the Sun, but also sometimes far, they could sometimes be up all night. These planets were considered "above the Sun" or **superior**, farther than the Sun from the Earth.

Adding spheres for the planets would not be the final complication. Further observations showed that the problem of the planets was not yet solved. When observed day after day, a planet's motion relative to the stars appears to be eastward, but at a specific position in its orbit, each planet would stop its eastward motion, move west for a while, and then resume its eastward motion. Any model attempting to explain the motions in the sky would have to account for this **retrograde motion**.

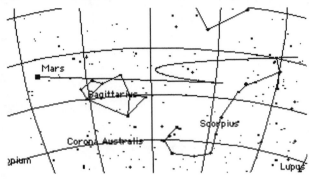

Figure 6.2 Retrograde motion of Mars during the summer of 2001

Retrograde motion was explained in the geocentric system with a device known as an **epicycle**. A planet was not thought to be directly in orbit of the Earth, but rather on a circular path called an epicycle. The center of the epicycle was in turn on a path called the **deferent**, which was in orbit of the Earth. The two circular motions occurring simultaneously would result in retrograde loops.

The periods of the planet's motion around the epicycle and the center of the epicycle's motion around the deferent were set in order to give accurate predictions of when the retrogrades would occur and how long they would last. It was also necessary to "fix" the epicycles of the inferior planets to a line between the Earth and the Sun so they would always stay close to the Sun.

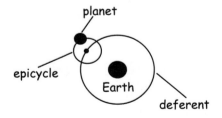

Figure 6.3 An epicycle and deferent

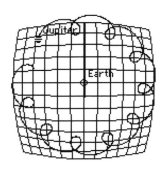

Figure 6.4 The motion of Jupiter on an epicycle and deferent

Explaining retrograde motion was the last step in making the geocentric model a viable explanation of the cycles in the sky. At this point it can be called a scientific **theory**. A scientific theory is the best explanation we have of something based on the available data, in this case observations. A theory is really the highest status that can be achieved in science. Theories must be tested over and over again; there is no "final test" that makes a theory a "law." A theory can be overturned at any time by new observations, no matter how much success it has had. If new observations show that the theory is not valid, it must be either discarded or revised. Note that the word theory has a different meaning in science than in any other context. In criminal justice, for instance, a theory about a crime is just a guess that must be proven before a conviction can occur. In science, as mentioned above, a theory is the best explanation we have.

The term "law" is often used as a colloquialism for a "well-tested" theory. The currently accepted use of the term "law" in science is for an observed relationship between experimental data. In **Chapter 8** we will see an example of a law under this definition.

THE SCIENTIFIC METHOD

Notice the process that occurred as the ancient Greek astronomers developed the geocentric system. Observations were made, followed by attempts at explanations. The explanations were then tested by further observations. When the further observation revealed that the current explanations were inadequate, the explanations were revised. The process of explanation, observation, and revision was repeated as many times as necessary until all observations were satisfactorily explained. At this point the explanations become a theory.

This process of observation, explanation, and revision is referred to as the **scientific method**. Science is NOT just a body of knowledge; it is rather the repeated process of observation, explanation, and revision by which the knowledge is attained. Many well-established theories have been successfully tested so many times that they may seem like a collection of facts, but to get to that point the process of science was applied time after time. Many theories that are less firmly established currently seem more like "works in progress." The questions about them that still need to be answered give us a chance to see the scientific process in action. Someday the ideas may be widely accepted, or possibly they will not be and could be replaced by new ideas. Only repeated application of the scientific process over time will tell.

It is important to realize that the ancient Greek astronomers were not following a "cookbook" set of instructions on how to apply the scientific process. Rather, in their attempt to understand what was going on around them, they invented it. This process of observation, followed by explanation, further observation to test the explanation, then revision repeated until all observations are explained is the way science has been done ever since.

Examination of the geocentric or **Ptolemaic system**, as it was also known after the publication of *The Almagest*, shows that it had all the features of a good scientific theory. The model worked, it included explanations for all observed motions, and it could be used to make accurate predictions. If you wanted to know where an object would be at a later date, the geocentric model could be used to accurately predict its position. The explanations also had a common-sense aspect to them. The motions that were observed, a rotating sky, and revolutions of the Sun,

Moon, and planets around the Earth were considered by many to be what was really happening. Despite the complexities of extra spheres for the planets and epicycles for their retrograde motions, this "common-sense" quality allowed the model to be perceived as fairly simple.

These features, accuracy and simplicity, were the major reasons that the Ptolemaic system was accepted for so long. Another reason will be discussed in the next chapter.

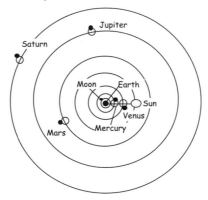

Figure 6.5 The geocentric or Ptolemaic system

Exercises

1.   A concept-map or a flow-chart is a device used to show the connections between the different steps in a process. On the next page is a sample concept-map of what a college student's typical day may be like.

The possible steps in the process of science are: *observation, explanation, testing, revision, abandon* and *theory*. Start by writing down the word observation and circling it. Then draw a line to another circle and put the next step in the process in that circle. Continue in this manner until you have used every step in the process and shown all of the connections between them.

You should use the sample below as a guide, but keep in mind that it is probably a MORE complex map than the one you will construct.

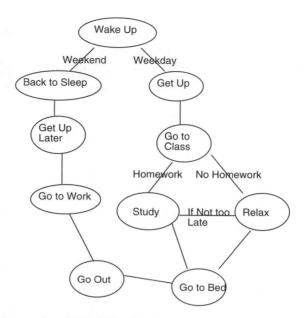

A concept-map of a (simplified) day in the life of college student

Construct your concept-map of the scientific process in the space below.

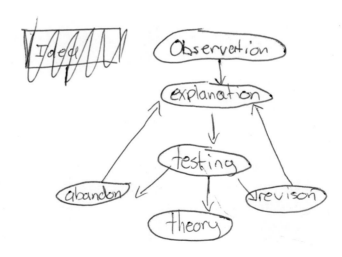

59

2.   Give an example from everyday life of the application of the scientific process. It can be about anything as long as you cite an example that includes observation and explanation followed by testing and, if necessary, revision of the explanation.

Questions

1.   Name two observations that the ancient Greeks could have made that would have been logical reasons for their belief that the Earth was spherical in shape.

2.   Name two observations that the ancient Greeks could have made that would have been logical reasons for placing the Moon's sphere closer to Earth's than the Sun's.

3.   What observation was most likely the reason for the order of the planets in the geocentric system? What observation that would have been logical to try and use would have turned out NOT to work?

Chapter Seven

THE HELIOCENTRIC
SYSTEM

ANOTHER MODEL

Despite the success of the geocentric model, another model did exist, not only prior to the time of Ptolemy but also prior to Aristotle. **Aristarchus** of Samos proposed a heliocentric model, where the Sun was central and the Earth was just one of the planets in orbit. He believed this because of calculations he made that showed that the Sun was much further away than the Moon. Since the two objects appeared about the same size in sky, this meant that the Sun must have been much larger. His calculations in fact showed that the Sun was much bigger than the Earth as well, so it made more sense to him that the Sun be central rather than the Earth.

Not too much else is known about why Aristarchus thought what he did because his book about his work was one of many lost in the burning of the Alexandria Library. Much of the knowledge of classical civilization was kept in this great storehouse, which was destroyed in a raid of conquest on Alexandria toward the end of the classical period, around 400 AD.

THE DARK AGES

Ptolemy's Almagest was one of the few works that did survive the civilization that produced it. Ptolemy actually mentions Aristarchus' heliocentric idea and concedes that attributing the observed motions in the sky to rotation and revolution of the Earth rather than the sky and the Sun is a viable explanation, but too complex. He felt that since everything could be explained in terms of the motions we actually see occurring, there was no reason to say they just appeared to be happening because of motions of the Earth, especially when there was no convincing evidence that the Earth moved. Recall Occam's Razor.

With the Christianization of the Romans, around 400 AD, the period of classical civilization ended and the period of history in western civilization known as the Dark Ages began. During this period of approximately 1,000 years, the advancement of knowledge nearly ground to a halt. Much of what was known during the classical period about agriculture, architecture, medicine, and many other things, not to mention science, was forgotten. Famine was widespread; people were dying of once curable diseases and once controllable plagues. Feudalism was the predominant form of government. People pledged their loyalty to a lord or baron in order to work a small part of his manor and exchanged what little spare crops they produced for protection. Petty feuds between neighboring manors were common. When people are worried about their health and whether they will have a place to live, food to eat, or protection from enemies, they tend not to worry about lofty pursuits like literature, the arts, or science, let alone specifically astronomy.

THE CHURCH AND ASTRONOMY

Throughout the Dark Ages, most flirtations with science were dismissed as mysticism or witchcraft, especially by officials in the Catholic or Christian Church, which had become very powerful (the terms "Christian" and "Catholic" were synonymous until the Reformation in the 1500s). By the thirteenth century, as living conditions improved, interest in science did not seem like it was going to stop. By then the Catholic Church, which was the governing body of western civilization, had changed views. Officials decided that if science could not be dismissed, it had to be controlled. St. Thomas Aquinas was a leader in this point of view. Aquinas was a theologian. Theology is the study of religion as an academic subject, independent of belief. He was one of the first "logical" theologians, attempting to study religion with the objectivity of the scientific method. This allowed Aquinas to understand the power of the scientific process. He was instrumental in getting the Church to set a policy on science. As a result of his efforts, Ptolemy's Almagest was adopted as the Church's doctrine on astronomy.

As discussed in the last chapter, the Ptolemaic system was a good scientific theory. It explained observations in a common sense-oriented fashion and it could be used to accurately predict the positions of objects. Now it had another advantage, it was the law. Much of the theory's staying power was derived from the feature that made it so easy for Christians to embrace it as dogma, Earth's central position. According to the theory, Earth was the center of the Universe, and since people were dominant on the Earth, the geocentric system gave people the same feeling of importance afforded to them by the Bible. Studying any other astronomical theory was breaking the laws of the church, a crime called heresy, which was at that time punishable by death.

THE HELIOCENTRIC SYSTEM

The heliocentric system resurfaced in the 1500s with the Polish Cleric Nicholas **Copernicus**. It is not clear whether he "reintroduced" it when he learned of Aristarchus' ideas or if he rediscovered it himself. Either way, being a churchman, he knew what could happen if he publicly championed the

Figure 7.1 Copernicus

Copernican system, as it came to be known, so he only circulated his ideas among trusted friends.

As seen in **Chapter 3**, the heliocentric system explains the daily motion of the sky with a daily rotation of the Earth and the annual motion of the Sun through the stars by an annual revolution of the Earth around the Sun. The seasons are explained by attributing the 23 1/2° angle between the celestial equator and the ecliptic to a tilt of the Earth's rotational axis. The tilt is relative to a line perpendicular to the line between the Earth and the Sun. As admitted even by Ptolemy, these explanations work as well as those in the geocentric system, but they are more complex because they assume that the Earth moves, something for which there was no compelling evidence.

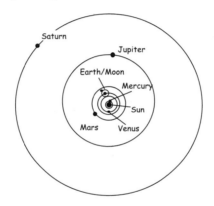

Figure 7.2 The Copernican system

The Copernican system even had an explanation for retrograde motion. According to the theory, when one planet passes another in orbit around the Sun, both planets will observe the other in retrograde motion. The outer planet will observe the inner retrograding after it passes and goes around the far side of the Sun. The inner planet observes something similar to a faster car passing a slower one on the highway. When you look in the rear-view mirror (which hopefully you do) the car you have passed will appear to be moving backward. Try it. This explanation will be demonstrated in an exercise at the end of this chapter.

The Copernican system seemed to work and it could be argued that the explanation for retrograde motion was simpler than epicycles, even though it still involved the Earth moving and the observed motion (like all observed motions in the heliocentric system) was considered an apparent motion, not an actual motion. However, when it came to making predictions of the positions of objects, the Copernican system, for reasons that will be seen in the next chapter, did not work as well as the Ptolemaic system. Most did not feel that there was a compelling reason to risk the penalties of scientific heresy by supporting it.

Copernicus did not publish his work until he was on his deathbed in 1543. His description of the heliocentric system was called *De Revolutionibus*. The term revolution does not refer to the Earth revolving around the Sun, but rather a revolution or change in scientific thinking. Copernicus died believing that his work would cause the world to change its view from a geocentric cosmos to a heliocentric one. He never knew it, but he was right. The events following his death, the over 100-year period it took to make this transition, are today referred to as the Copernican Revolution.

REACTION

Even though the heliocentric system did not seem to pose an immediate threat to the Church-supported geocentric system, the Church reacted swiftly to the threat to its authority. Not only the Catholic Church, but also the Lutheran Church, which had recently broken away from the Catholics, put *De Revolutionibus* on a list of forbidden books. There is currently some controversy as to how violent a verbal attack Martin Luther actually made on the late Copernicus, but the work was nevertheless banned. Many became naturally curious about what was in the book that they were not supposed to see and some sought to attain a copy. In this way, the churches may have actually promoted the system they were trying to suppress.

A REASON FOR A CLOSER LOOK

By the late 1500s there were many debates about which system was better. The geocentric system had tradition and the law on its side and it actually worked better than the heliocentric system when it came to predictions. There was, however, one feature of the heliocentric system that did not exist in the geocentric system. This was a method for determining the distances between the Sun and the planets relative to the Earth's distance.

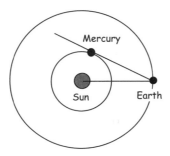

Figure 7.3 Copernicus' method for determining the distance of an inferior planet from the Sun

The only observation needed is the angle of the greatest elongation of the planet, the farthest the planet is ever seen from the Sun, in the case of Mercury, 28°, and Venus, 46°. On paper, a line is drawn to represent the distance between the Earth and Sun, and a compass is used to draw the Earth's orbit centering on the Sun. The angle of greatest elongation is measured with a protractor from the line between the Earth and the Sun, and then a line is drawn from the Earth at this angle in the direction of the planet. A circle centering on the Sun can then be drawn with the compass that just touches the line that represents the planet's greatest elongation. This will be the planet's orbit.

The actual distance between the Earth and the Sun was not known in Copernicus' time; it was just referred to as 1 Astronomical Unit or AU for short. The distance from the Sun to the planet's orbit is measured with a ruler and divided by the distance measured from the Sun to the Earth. This will give the planet's distance from the Sun in astronomical units.

The method for determining the distances for superior planets is more complex, but works just as well. Figure 7.4 compares Copernicus' distances to modern accepted values.

It is important to understand that no one in Copernicus' time actually knew whether or not the values were "correct." However, just the fact that they could be determined was an interesting feature of the Copernican system. This was a feature interesting enough to warrant further study by astronomers after the time of Copernicus.

Planet	Copernicus' Value (AU)	Modern Value (AU)
Mercury	.38	0.387
Venus	.72	0.723
Mars	1.52	1.524
Jupiter	5.2	5.203
Saturn	9.2	9.539

Figure 7.4 Copernicus' values for the distances from the Sun to each planet compared to modern values

Exercises

1. The picture on the next page shows Earth passing Mars as both orbit the Sun. Note the line drawn from the first position of the Earth through the first position of Mars, and on to the background on the right. The point where the line ends up on the background has been numbered 1. Use a ruler to draw a line for each successive set of corresponding positions of Earth and Mars. After you have drawn lines through the two planets and numbered the points on the background for all seven remaining positions connect the points on the background *in numerical order*. The result should show Mars' apparent path through the background as observed from Earth.

Mars during its 2003 opposition. Taken at Edinboro University of Pennsylvania-Maize Sunfire Observatory. Used with Permission of the Director.

Chapter Eight

THE COPERNICAN
REVOLUTION
PART 1

TYCHO

One of the first astronomers to make an attempt at determining which system worked better was the Danish nobleman **Tycho** Brahe. Tycho (history has remembered him by his first name) rented the island of Hveen from the King of Denmark and constructed one of western civilization's first observatories. Tycho had no telescopes at his observatory; his observations were strictly naked eye. He made observations for many years and kept careful track of the positions of planets, mostly Mars, as they moved through the sky. Tycho did not believe that the Earth moved, and therefore, felt that his observations would support the geocentric system. The reason that Tycho did not believe that the Earth moved was because of the absence of stellar parallax in his observations.

Parallax is the apparent shift in the position of an object due to a change in the observer's point of view. It is easy to demonstrate parallax. Hold your thumb out at arm's length and without moving it either close one eye and shake your head or keep your head still and alternately open and close each of your eyes. You will notice your thumb appearing to move even though you know that it is not. The apparent motion of your thumb is caused by parallax.

Although some thought that all stars were fixed to a giant sphere, others made the reasonable assumption that the brighter ones were closer than the dimmer ones. If the Earth was indeed in orbit around the

Sun, some of the brighter stars should show a parallax shift over the six months it would take the Earth to move from one side of the Sun to the other. Tycho never observed this stellar parallax. To him, this was convincing evidence that the Earth did NOT move and was the reason for his support of the geocentric system.

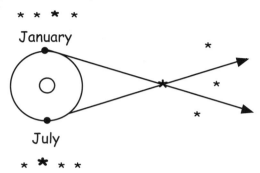

Figure 8.1 Stellar parallax due to a moving Earth

In hindsight, since the heliocentric system is accepted today, it is important to ask why Tycho failed to detect stellar parallax. If you cannot answer now, think about it for a while and try again at the end of this chapter.

Tycho was an example of what in science is called an experimentalist. He was the greatest observational genius of his time and had the best astronomical data that the world had yet seen. However, he was not a theorist. A theorist is a scientist who can take data and use it to explain what is being observed. It is possible to do both, but most scientists are better at being one or the other. Tycho developed a system that was somewhat of a cross between the two existing systems, but even he did not take the system seriously. He needed the help of a theorist.

Figure 8.2 Tycho and his observatory

KEPLER

Johannes **Kepler**, a German mathematician teaching in Graz, Austria, was that theorist. Tycho was in Austria by this time as well, working as the royal mathematician for the Archduke in Prague, circa 1600. Tycho had an argument with the King of Denmark. Being very arrogant, he refused to back down and was kicked off his island and out of Denmark. Kepler had been working on his own version of the heliocentric system based on geometry when Tycho invited him to come to Prague. Kepler was reluctant at first, but had no choice when Graz was taken over by Catholics preceding the 30-Years' War following the Reformation. His school was closed and he, being a devout Lutheran, was exiled.

Kepler could not make his geometric model work and knew Tycho had better data than he had access to. Unfortunately, Tycho would not give Kepler full access to the data, perhaps being paranoid that Kepler would take all the credit for anything that was learned from the data. Just when Kepler was ready to leave in 1601, Tycho died of an exploded bladder after drinking too much at a party he had thrown for the Archduke. Kepler inherited Tycho's job and the data and could finally get to work.

Figure 8.3 Kepler and his geometric model

THE LAWS OF PLANETARY MOTION

Kepler worked for 8 years trying to fit Tycho's data to his geometric model. His idea was that the 5 spaces between the spheres that held up the 6 planets (including Earth) were based on nesting the Pythagorean Solids one inside another. The ancient Greek mathematician Pythagoras had discovered that there were only 5 three-dimensional shapes that had identical sides. When his model did not work, Kepler assumed it was Copernicus' calculations, mentioned in **Chapter 7**, of the distance between the planets and the Sun, and not his model, that were the problem. He believed that Tycho's data would show that he was correct.

According to Tycho's data, the orbit of Mars was within eight angular or arc minutes (1'=1/60 of 1°) of being circular. However, Tycho's data was accurate to 1 minute so Kepler realized that the path was actually elliptical, not circular. This was the beginning of Kepler's derivation of what would later be called the **Laws of Planetary Motion**. Kepler derived three relationships from Tycho's data:

Kepler's Laws of Planetary Motion

1. Planets orbit the Sun in elliptical paths with the Sun at one of the focus points.
2. A line between the Sun and a planet sweeps out equal areas in equal amounts of time.

3.   The square of the period of a planet's orbit is equal to the cube of its average distance from the Sun.

An **ellipse** is a set of points in a plane whose summed distance from two points in the same plane is equal. The two points are called foci, the plural of **focus**. An ellipse can be drawn by putting a loop of string around two tacks holding a sheet of paper to a piece of cardboard then drawing everywhere on the paper where the string can be pulled taut by a pencil.

In the **Law of Elliptical Orbits** the Sun is located at one focus, so the planet's distance from the Sun varies throughout its orbit. The point closest to the Sun is called **perihelion**. The farthest point is called **aphelion**. A line drawn from one side of the ellipse through the foci to the other side is called the major axis. Half this distance, the **semimajor axis**, is also the planet's average distance from the Sun.

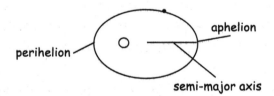

Figure 8.4 A Planet in an elliptical orbit around the Sun

The **eccentricity** of an ellipse is a measurement of how far it deviates from being a circle. An ellipse with zero eccentricity is a circle. The higher the eccentricity, the more elliptical the orbit. Most planetary orbits have very low eccentricity. Earth's average distance from the Sun, known as one **astronomical unit**, **AU** for short, is about 93 million (93,000,000) miles. At perihelion, in January, it is only about 1.5 million (1,500,000) miles closer than that. At aphelion, in July, it is only that much farther. One and a half million miles may sound like a lot, but it is less than 2% of the Earth's average orbital distance. Even though these eccentricities are small, they are significant. Copernicus' use of circles and not ellipses was the reason (mentioned in **Chapter 7**) that, when it came to predicting the positions of objects, his model never worked quite

75

as well as Ptolemy's. The eccentricities of the orbits of all the planets are included in the planetary data in **Appendix 4**.

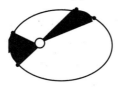

Figure 8.5 The Law of Equal Areas

The **Law of Equal Areas**, as it is called, has an interesting consequence (see Figure 8.5). When the planet is close to the Sun, the line between the Sun and the planet "sweeps out" a short, fat wedge. When the planet is farther from the Sun, it sweeps out a long, thin one. Since the equal areas are swept out in the same amount of time, the planet must be moving faster in its orbit when close to the Sun and slower when farther away. This consequence led Kepler to believe that the Sun had an effect on the planet that was stronger when the planet was closer to the Sun. This foreshadows the idea of gravity that will be discussed in **Chapter 10**.

The third law is often referred to as the **Harmonic Law**. Kepler believed in what he called "cosmic harmony." He believed that the planets sung as the speed, and thus the frequency, of their orbits varied (the frequency of a sound wave is what is responsible for the pitch of the sound that we hear). He felt that there should be a simple relationship between the **period**, p, the amount of time a planet takes to complete its orbit, and the average distance between the planet and the Sun, the semimajor axis, a, of the elliptical orbit. It took him 10 more years after he formulated the first two laws in 1609, but he finally determined the relatively simple relationship $p^2=a^3$.

It is important to understand that Kepler's laws are **empirical**. This means that they are derived from data, in this case Tycho's observations. He had no underlying physical reason for the relationships; he did not know why they were true, just that the observations showed that they were. This also means that Kepler's laws are an example of the currently

Chapter Nine

THE COPERNICAN REVOLUTION PART 2

GALILEO

When Kepler recognized that as a consequence of the Law of Equal Areas, planets move faster when closer to the Sun, he thought that there must be an effect that the Sun has on the planets that varied with distance. He realized that further investigation of this effect lied in the realm of physics. Kepler wrote to a physicist that he knew of in Italy and asked him for help in pursuing the idea. This physicist was in the process of conducting experiments that would change the world's view of physics, but he wasn't interested in Kepler's Laws. This doesn't mean he was not interested in astronomy. He was, in fact, an avid supporter of the Copernican system, like Kepler. However, he believed that astronomy should be studied in another way, with a telescope.

TELESCOPES

Galileo Galilee did not invent the telescope, but he was the first known to use it for astronomy. In 1609, the same year that Kepler came out with the first two laws of planetary motion, Galileo began making controversial observations. He discovered craters and mountains on the Moon and spots on the Sun. Although not having anything to do with whether or not the Earth orbits the Sun, these observations suggested imperfections in the "heavenly spheres." He was brought in front of the Inquisition for publicizing these observations and warned not to continue on his heretical course. He then turned his telescope to the brighter planets. He discovered 4 stars, which he realized were moons orbiting Jupiter and that Venus went through phases very similar to those of our Moon.

Jupiter having moons does not challenge the Earth's central position, but it does show that not everything orbits the Earth. However, the phases of Venus being the same as the Moon's could not occur in the geocentric system.

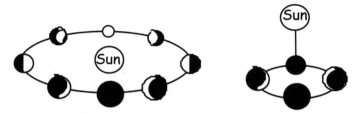

Figure 9.1 Phases of Venus in the heliocentric and geocentric systems; the phases in the heliocentric system (left) were what Galileo observed.

Despite the earlier warnings, Galileo published his observations, publicly declaring his support of the Copernican system. He was arrested, tried, and convicted of scientific heresy. He was sentenced to be burned at the stake, but was offered the chance to have his sentence commuted to life under house arrest if he would publicly recant, or take back, his support of the Copernican system. He apparently was not prepared to become a martyr to his cause, because he did recant and spent the rest of his life under house arrest.

GALILEO AND THE CHURCH

Galileo's case brings up the potential for conflict between religion and science. It is easy to cast the Church in a negative role, standing in the way of scientific progress, but it was more than religion alone that was responsible for Galileo's fate. Historians have spent much time analyzing this case, so there is no point in rehashing all the details, but a few comments are appropriate. Besides the fact that he was violating

Church dogma by supporting the heliocentric system, his condemnation was also partly political and personal. It was political, because he was rebelling against governmental authority, at that time the Church. It was personal because in his arrogance Galileo made many enemies, including the Pope, a former friend whom he mocked and insulted in his publication, *A Dialog on Two Chief World Systems*, as representing an ignorant establishment. More recently, in 1979, Pope John Paul II began proceedings that by 1993 cleared Galileo of the charge of scientific heresy.

At the time of his house arrest, Galileo was actually far from finished. He continued to make telescopic observations and wrote a book about his physics experiments that will be discussed in the next chapter. The work was every bit as "heretical" as his work in astronomy, but seemed to be unnoticed by the Church.

Figure 9.2 Galileo

RELIGION AND SCIENCE

A perspective that sometimes helps those who are struggling to reconcile parts of their religious faith with certain aspects of science is that while science is based on testability, religion is based on faith. If someone has a faith, if they *believe* in a religion, it does not *need* to be proven to them. Someone who requires proof of a religion does *not* have faith. Science, however, does not work this way. As you now know, scientific theories *must* be continually tested; they *cannot* be taken on faith. Subscribing to a faith is a decision of the individual, no one can tell you what to believe. With a scientific theory, the question is if the experimental and observational evidence supports the theory. This can make comparing the two disciplines somewhat like comparing the proverbial "apples and oranges." Religion and science are based on different tenants, religion on belief and science on testability, so for many there is not any conflict. This view has been adopted by many established churches as their policy on science.[1]

NEWTON

To find the next great scientist in our story we must go to England. Isaac Newton was studying at Cambridge in 1662 when the "Black Death" broke out in London. Newton retreated to a cabin in the country provided by his uncle, who was also paying for his education. It was there that he did some of his most profound thinking and produced some of his greatest work.

Figure 9.3 Newton

Newton was the first to describe physical theories in the language of mathematics. The idea of using the tool of mathematics to study science was not new; it went back, as did so many other things, to ancient Greece. However, Newton actually used mathematical equations to apply his theories and make predictions that could then be tested by observation or experiment. At the time there was no form of mathematics suited to the task, so he invented calculus, the highest form of mathematics. Tools are usually invented when the need for them arises.

Newton's **Law of Universal Gravitation** is a mathematically-based theory that describes **Gravity** as a force between masses that pulls them together. The force varies proportionally with mass, more mass means more gravity, and inversely with distance, the greater the distance between objects the weaker the gravity between them. The drop-off in gravitational force with distance is quite drastic; Newton determined that the force was proportional to the inverse of the square of the distance. So, for example, doubling the distance between two masses decreased the gravitational pull by a factor of 4.

---

[1] The author's viewpoint on this issue is expressed in; M. LoPresto "*Dealing with Conflicts between Religion and Science in Introductory Astronomy,*" **Mercury**, Nov.–Dec., 1999, pp. 36–37.

Chapter Ten

SOME PHYSICS

WHAT IS PHYSICS?

Besides their contributions to astronomy, Galileo and Newton also made huge contributions to physics. Physics, coming from the word "physical" is the study of the physical universe. It is our attempt at understanding the reasons that matter and energy behave the way they do. When studying astronomy, it is important not to shy away from physics, especially the areas that are specifically relevant to astronomy.

ARISTOTLE'S PHYSICS

Aristotle, first mentioned in **Chapter 6**, had some ideas about the world around him. He felt that motion was a result of objects "wanting" to return to their "natural element." Aristotle believed that everything in the world was part of one of the four "elements," the spheres of *Earth, Water, Air,* and *Fire*. When solid objects fell, they were trying to return to *Earth*, when rain fell into the ocean or fell on land then ran off to lakes and rivers, it was returning to *Water*. Steam floats upward towards *Air*, lightning and flames from a campfire shoot upward to the sphere of *Fire*.

When thinking about the difference between the motion of a falling rock and a falling leaf, he assumed that the rock fell faster because it was heavier than the leaf. Aristotle described the *mass* of an object as an "eagerness" to return to its element. The rock, being more massive, was more eager than the leaf.

GALILEO'S EXPERIMENTS

Just like geocentric astronomy, Aristotle's ideas about motion made sense and agreed with observation. Also as with geocentric astronomy, it was not until the Renaissance that they were seriously challenged. Galileo's greatest contribution to physics and perhaps science in general was the idea of experimentation. Experiments are basically observations of a situation set up by the observer. The ancient Greek thinker Archimedes was known for doing experiments, but it was Galileo that reintroduced experiments to the western world and essentially began what we now call the "Age of Modern Science."

Galileo probably never actually did the experiment for which he is most famous. Legend has him dropping two cannonballs of different mass off the Leaning Tower of Pisa and observing them falling at the same rate, contrary to what Aristotle would have predicted. He did do other experiments, however, which were leaning (no pun intended) in the same direction, such as rolling masses down inclined planes.

Aristotle attributed motion to the object; Galileo had the idea that changes in motion were caused by external forces. Instead of a rock or cannonball wanting to return to Earth, he suggested that they were being pulled to Earth by a **force** (later identified by Newton as gravity). Also, just as objects could be pulled down by a force, they could also be pushed up, as in the case of the leaf's fall being cushioned and slowed down by the air. Instead of mass being an "eagerness" of the object, Galileo described it as a resistance to forces. The more massive cannonball is more resistant to the Earth's pull and, therefore, only falls at the same rate as the less resistant, smaller one.

Galileo published these ideas after his house arrest. As mentioned earlier, they were every bit as "heretical" as his astronomical observations, but he was not prosecuted (or persecuted) further.

NEWTON AGAIN

Just as he "finished" the job in astronomy by theoretically verifying Kepler's Laws of Planetary Motion with the Universal Law of Gravitation, Newton finished the work of Galileo as well. Born in 1642, the same year Galileo died, Newton verified Galileo's observations with what are known simply as the **Laws of Motion**. They are basically mathematical applications of the ideas Galileo had experimented with. Objects stay in their current state of motion, whether at rest or in motion, and the only way to change the state of motion is through the application of force. This resistance to motion is known as **inertia**. Mass is a measurement of inertia. When Newton's laws of motion are mathematically applied to the situation of objects falling to Earth, the result is that they fall at the same rate regardless of their mass. For instance, if one object is twice as massive as another, it will be pulled to Earth by twice as much gravity, but it will also have twice as much resistance to that pull, so the objects will fall at the same rate, which was exactly what Galileo observed.

Another important point in the Laws of Motion is that forces always come in pairs. A cup sitting on a table remains at rest because an upward force from the table balances the downward pull of gravity. The two forces are an "action-reaction" pair. A relevant example of this is that the Earth pulls on the Sun with every bit as much gravity as the Sun pulls on the Earth. The Earth ends up orbiting the Sun and not vice versa because the more massive Sun has much more inertia than the less massive Earth. This is why the Moon orbits the Earth and also why a cannonball falls to Earth rather than pulling the Earth up to it!

CONTRIBUTIONS TO SCIENCE

Galileo and Newton's contributions to astronomy are obvious. Galileo introduced the telescope and made many discoveries, including Venus going through a cycle of phases similar to the Moon's, a cycle that would not be possible if the Sun orbited the Earth. Newton, with Universal Gravitation, was able to theoretically verify Kepler's Laws, which came from Tycho's observations, effectively finishing the Copernican Revolution.

Just as Tycho and Kepler had, Galileo and Newton also made monumental contributions to science in general. Galileo reintroduced experimentation to western civilization, and Newton mathematically applied physical theories to verify observations and experiments, refining the scientific method into the process it is today.

THEORIES AND LAWS IN SCIENCE

Newton's Laws of Motion and Universal Gravitation are well-tested theories. However, as mentioned before, they can never pass a "final test" to become "law" and can always be overturned if new observations prove contradictory. Also as mentioned previously, an example of the currently accepted use of the word "law" in science is in Kepler's Laws of Planetary Motion. Laws are considered relationships between data based on observations.

Just as the very successful geocentric system was eventually overturned by the Copernican Revolution, Newton's Law of Universal Gravitation "driving the final nail..." Newton's Universal Gravitation suffered a similar fate at the hands of Einstein's **General Theory of Relativity**. General Relativity is a new way to look at gravity, considering it a curve in the fabric of space caused by mass rather than a force between masses. Its formulation was necessitated by the fact that in cases of extreme gravity, Universal Gravitation does not provide an adequate description of what is observed to happen.

DO WE REALLY KNOW IF THE EARTH MOVES?

Stellar parallax, first discussed in **Chapter 8**, suggests that the Earth moves. This was not, however, a factor in the Copernican Revolution because the stars are so far away that they do not show enough parallax

90

to be detected without a telescope, in fact without a powerful telescope. It has only been a little more than 100 years since stellar parallaxes have been detectable.

The best way to "observe" the motion of the Earth is with a pendulum. A pendulum is simply a mass attached to a string. If a pendulum is pulled back and allowed to swing long enough without losing its energy to friction, the direction of its swing will appear to change as a result of the Earth rotating under it. This is because, not being directly attached to the Earth, the pendulum's mass will not rotate with the Earth and its inertia will keep it swinging in the same direction. Pendulums of this type can be found in many natural history and science museums.

PHYSICS AND ASTRONOMY

Many areas of physics are useful in astronomy; this is the reason that the development of the two disciplines have paralleled each other throughout history. We have already seen how the study of motion, known as **Mechanics**, which includes gravity, grew with its application to planetary motion. As you will see in **Chapter 11**, light travels through space as an electromagnetic wave, a combination of varying electric and magnetic fields. This and the fact that astronomers use lenses and mirrors in telescopes to collect, and other instruments to analyze, light make both **Electromagnetism** and **Optics**, the study of light, important to them. **Thermodynamics**, the study of heat and temperature, can be used in determining the Temperature of the sources of light and how energy is transferred through space and within stars.

The above are the four major areas of what is known as "*Classical Physics.*" "*Modern Physics,*" as it is called, involves **Atomic Physics**, the study of the atom. Understanding atoms is necessary for astronomers to determine the composition of distant objects from the light they emit and how the light itself is created. **Nuclear Physics** is the study of the nucleus of the atom. Nuclear energy provides the fuel that allows stars to shine and give off heat. **Elementary Particle Physics** is the study of fundamental particles and the interactions between them. Particle Physics is also important in understanding the nuclear energy that keeps a star shining as well as the remnants left at the end of its life cycle. It is also an integral part of Cosmology, the study of the Universe as a whole. Studying how particles and interactions change as temperatures get higher and higher allows astronomers to theorize about how the Universe itself began.

Atomic, Nuclear, and Particle physics are all within the realm of the very small, the microscopic universe. **Quantum Theory**, the idea that particles can be described as waves and waves can be described as particles, is the theory that governs and unifies the microscopic universe. For instance, light can be thought of as an electromagnetic wave when it travels, but as a particle called a photon when it is created. The macroscopic universe is governed by **General Relativity**, Einstein's theory of gravity. **Special Relativity**, which explains how matter and energy behave at speeds approaching that of light and that light is a "cosmic speed-limit" is a part of the more general theory.

The ultimate goal of physics is to discover a **Grand Unified Field Theory**, or **Theory of Everything**, which explains both the physics of the very small and the very large at the same time. There are partial theories of this type, combining some, but not all, observed interactions, but a complete description, which would provide a fundamental understanding of the Universe, has as of yet eluded us.

Chapter Eleven

OTHER "GREAT MOMENTS" IN SCIENCE

As mentioned in previous chapters, the greatest moment in science is when theory and experiment achieve the same result, thus verifying one another. Newton first achieved this when his theories on gravitation and motion verified the observations of both Kepler and Galileo. This chapter highlights a few more examples of these "great moments" that are relevant to astronomy.

THE SPEED AND NATURE OF LIGHT

In the late 1800s physicist James Clerk Maxwell was studying the nature of electricity and magnetism. He took the four basic equations that described the phenomena, the laws of Faraday, Ampere and Gauss (Gauss had two, one each for electricity and magnetism) and solved them simultaneously. You may have simultaneously solved algebraic equations at some point. Solving differential equations simultaneously is similar, but a little more difficult. The solution he obtained was the equation for a wave. The speed of these "electromagnetic" waves was also part of the solution. It was a function of the basic constants of electricity and magnetism and turned out to equal about 300,000,000 m/s (3×10^8 m/s in scientific notation). This is, of course, the speed of light. The four equations of electromagnetism have been referred to as *Maxwell's Equations* ever since.

The existence of electromagnetic waves was also verified experimentally by Heinrich Hertz, who actually produced them with vibrations of an electrified antenna. Because of his contribution to the physics of waves, the unit of frequency is named the hertz.

Later, physicist Alfred Michelson was performing precise experiments to measure the speed of light waves and was obtaining the exact same result as Maxwell's theoretical speed for electromagnetic waves. Michelson was the last in a line of scientists who attempted measuring the speed of light, Galileo being the first. Another was German astronomer Ole Roemer. He used a method involving the moons of Jupiter *that Galileo discovered*. However, Michelson's experiments were the most precise and the first that could be compared with theoretical predictions.

With Maxwell's theory and the experiments of Hertz and Michelson, both the speed and nature of light were known. This is very important to astronomy, since almost all the information used by astronomers comes in various forms of light.

Figure 11.1 Maxwell

ATOMIC STRUCTURE

One of the properties of astronomical objects that can be determined by examining the light they give off is their composition. This can be done by separating the light from the object into the different components of its **spectrum**. The device used to do this is called a spectrometer, and the study of spectra is called **spectroscopy**. The operative element in a spectroscope is either a glass prism or a device known as a diffraction grating, many lines etched very close together into a piece of glass.

Spectroscopy has shown us that the observable universe is made chiefly of hydrogen. In the laboratory, heated hydrogen gas glows with a purplish pink color. When the light is observed through a spectroscope there are four visible lines in the spectrum: two violets, one blue-green, and one red. The color of light is controlled by the length of the wave. This is known, not surprisingly, as **wavelength** and can be readily measured using a spectroscope and the properties of waves. Different gases all have unique spectra. Since gases are made of atoms, that is where the light must be ultimately coming from, so the spectra are referred to as **atomic-spectra**. Each spectrum being different suggests that the structures of the atoms of which the different gases are made are also different.

A physicist named Jakob Balmer developed an empirical formula to calculate the wavelengths of the visible lines of hydrogen. These lines have been called the Balmer series ever since. The Balmer formula, as it is called, depends on a constant that was first determined by Johannes Rydberg and bears his name. In the study of atomic spectra, Balmer and Rydberg played roles very similar to those of Tycho and Kepler in the study of planetary motion. The Balmer formula and Rydberg constant had no physical meaning; they just accurately predicted the wavelengths of the hydrogen spectrum.

Danish physicist Niels Bohr played the "role" of Newton. He constructed a theoretical model of the hydrogen atom based on known principles of physics. From his model of the hydrogen atom he was able to derive the Balmer formula and the Rydberg constant. This meant that his theoretical model atom had the same spectrum as had been observed experimentally with gas made of real hydrogen atoms. The agreement of Bohr's model with observed spectra is the basis of our understanding of atomic structure.

Another great lesson to be learned about science here is how one question leads to another and how you never know what you will learn from your inquiries. The initial relevance of atomic spectra to astronomy is under–standing how the spectra of stars and galaxies can tell us their composition. Attaining this understanding also makes necessary an understanding of atomic structure and also the very creation, within the atoms, of light itself.

Figure 11.2 Bohr

THE EXPANDING UNIVERSE

The theory of general relativity, first mentioned in **Chapter 10**, was Albert Einstein's new way to look at gravity. He treated gravity as a curvature in the fabric of space caused by mass, rather than as a force between masses. In the 1930s when he applied his field equations, as they were called, to study the dynamics of the universe as a whole he came up with a remarkable result, that *the universe is expanding*. This result puzzled him, because he believed that, on a large scale, the universe should be static and unchanging. To "fix" this result he inserted a term he called the cosmological constant, which stopped his theoretical universe from expanding.

Around the same time, Edwin Hubble was making remarkable observations from Mt. Wilson in California. When examining the spectra of light from distant galaxies, he noticed that they were all **redshifted**. Red is the color of the longest visible wavelength of light. A shift toward the red end of the spectrum means that the waves coming from the source are being stretched out. This shift occurs when an object is moving away from the observer, leaving the waves behind. So, if a galaxy shows a redshift in its spectrum, it is moving away from us. Since all galaxies beyond those in our "local" cluster show a redshift, they are all moving away from us. If they are all moving away from us, the Universe must be expanding, just as Einstein had predicted.

Hubble had Einstein come out to California to see his data. When Einstein realized that Hubble's observations had verified his original theoretical prediction, he called the addition of the cosmological constant "the greatest blunder of his scientific career."

Figure 11.3 Hubble and Einstein

Interestingly, Einstein's cosmological constant, which is a repulsion force meant to balance the pull of gravity, has recently resurfaced as modern cosmologists have made new observations that seem to show that the rate of the Universe's expansion may be increasing.

Chapter Twelve

MOTIONS OF
THE PLANETS

The observed period of the Sun is a result of the Earth's revolution around it. One year, the amount of time required for one revolution, is Earth's sidereal period. The **sidereal period** for any planet is the time required for one orbit around the Sun. This is the period used in Kepler's third law.

A planet's **synodic period** is measured in reference to the Sun from Earth. It is the amount of time required, as viewed from Earth, for a planet to cycle through all possible configurations with the Sun.

The synodic period can be observed; the sidereal period, however, cannot be. It must be calculated from the Earth's sidereal period and the planet's observed synodic period. This is because a planet's orbit is centered on the Sun.

Figure 12.1 shows the synodic period of an inferior planet, Mercury or Venus. When a planet lies along a line that passes through both the Earth and the Sun, it is said to be in conjunction. **Inferior conjunction** is when the planet is "below" the Sun or between the Sun and Earth. **Superior conjunction** is when the planet is "above" the Sun or the Sun is between the planet and Earth.

A planet will be difficult to see when near either conjunction because of the brightness of the Sun. Since the planets do not lie exactly along the ecliptic, the Sun will not always eclipse them when they are in superior conjunction. If the Sun does eclipse them it is known as an **occultation**. If a planet at inferior conjunction passes in front of the Sun it is called a **transit**.

Note that this seems like a slightly different use of the term transit than in earlier chapters, but really it is not. In both cases "transit" refers to an object passing something, in this case, a planet passing in front of the Sun. Our earlier definition was an object crossing the meridian.

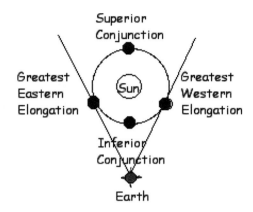

Figure 12.1 Configurations during the synodic period of an inferior planet

The apparent angle between an inferior planet and the Sun is called its **elongation**. Conjunctions are 0° elongation, and each planet has a greatest eastern and western elongation. Venus appears at most about 46° degrees from the Sun and Mercury at most about 28°. This is the position from which they can be best seen from Earth. You may recall that Copernicus used greatest elongations to determine the distance of Venus and Mercury from the Sun relative to the Earth's distance (see **Chapter 7**).

The configurations during the synodic period of a superior planet are shown in Figure 12.2. **Superior Conjunction** occurs when the sun is in between the planet and the Earth. When the Earth is in between the planet and the Sun, it is said to be at **opposition**, because the planet is opposite the Sun in the sky. **Quadrature** is when the planet appears 90° from the Sun in the sky.

Throughout its synodic period, a planet will cycle through various phases that can be observed from Earth. It will be left as questions at the end of the chapter to determine which phases occur in which configurations and to determine at what times the planets are visible from Earth. Another exercise at the end of the chapter involves constructing graphs of various observable properties of inferior and superior planets as seen from Earth.

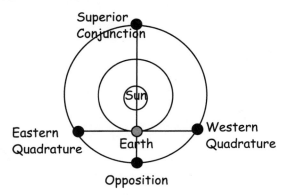

Figure 12.2 Configurations during the synodic period of a superior planet

THE DISCOVERY OF PLANETS

Astronomers have known the five planets visible to the unaided eye, Mercury, Venus, Mars, Jupiter, and Saturn, since ancient times. It was not until after the invention of the telescope that new ones could be discovered. In the late 1700s, German Astronomer Johann Titus devised a mathematical formula that matched the distance of each of the known planets from the Sun relative to Earth's fairly well. Starting with a series of numbers beginning first with 0, then 3, and then doubling 3 and each number after that, then adding 4 to each number you get, and finally dividing them all by 10 gives the values shown in the table in Figure 12.3. The idea was publicized by the then director of the Berlin Observatory, Johann Bode, and thus called the **Titus-Bode Law**. Interestingly enough, 2.8 AU, where there was no known planet, was a location where many, including Kepler, thought there probably was a planet that had not yet been discovered.

A Series of Numbers	Add 4	Divide by 10	Known Distance (in AU)	Planet
0	4	0.4	0.38	Mercury
3	7	0.7	0.72	Venus
6	10	1.0	1.0	Earth
12	16	1.6	1.52	Mars
24	28	2.8		
48	52	5.2	5.2	Jupiter
96	100	10.0	9.5	Saturn

Figure 12.3 The Titus-Bode Law

URANUS

Using the Titus-Bode law to calculate the position of the "next" planet after Saturn gives 19.6 AU. Several astronomers had searched the sky for a planet at this position, but nobody found anything until 1781. William Herschel was born in Germany but living in England. A musician by day and an amateur astronomer at night, he had built a large reflecting telescope. He was actually more interested in studying stars and in fact would eventually be known as the father of stellar astronomy. Over the course of observing the same part of the sky on several nights, he noticed an object that was not a point of light, but a hazy disk that, over several observations, appeared to be moving in front of the fixed background stars. This relative motion is the telltale sign of a planetary object. He had dismissed the object as a comet, but closer investigation showed that the object was in an orbit that was too circular for a comet. Comet orbits are very eccentric ellipses. Although the discovery was at first resisted by professional astronomers, probably due to jealousy of an amateur making a discovery that they did not, the quality of Herschel's telescope, and thus the validity of his observations, could not be denied. Soon he was credited with the first discovery of a planet since ancient times and before long received a royal appointment as a full-time astronomer. His sister Caroline, who started as his assistant, became a successful astronomer in her own right and received an appointment as well.

The discovery created an interesting problem that no one ever had before, what to name a new planet? Some wanted to name it after the discoverer. Herschel himself wanted to name it after his sponsor, the King of England. Whether Herschel and George would have been good names for a planet could be debated. However, it was decided that the ancient tradition of naming

Figure 12.4 Herschel

planets after mythological characters should be continued, and the new planet was named Uranus.

NEPTUNE

Uranus turned out to be 19.2 AU from the Sun, matching the Titus-Bode law as well as the other planets. At this distance it orbits the Sun once every 84 years, which coincidentally is also how long William Herschel lived. Needless to say, astronomers became excited about using the Titus-Bode law to find planets. This excitement intensified when Giuseppe Piazzi discovered Ceres, the largest member of the asteroid belt, in 1801 at 2.8 AU, what had been called the position of the missing planet. The next predicted position was 38.8 AU, however, by the mid-1800s, astronomers had another method for searching for planets.

Irregularities in the orbit of Uranus, first detected in 1821, suggested that there was another sizable object beyond Uranus exerting gravitational influence other than Jupiter or Saturn. In 1844 a young English mathematician, John Couch Adams, used these perturbations to predict the position of the planet at about 30 AU. All he needed now was time on a good telescope to search for it. He tried to convince British Astronomer Royal, George Airy, to get him time on a large telescope for the search, but Airy was difficult to contact, skeptical of Adams' idea, and did not help him.

A young French mathematician named Urbain Leverrier had made similar predictions to Adams'. He also had similar frustrations in getting telescope time. So he went to Germany in 1846 and asked Johann Galle, director of the Berlin Observatory, to take up the search. Galle found the object on the very first night.

It is not hard to imagine Adams' frustration when Leverrier and Galle published the results and were credited with the discovery of Neptune. Much controversy, intensified by nationalistic rivalry between England and France, followed. Having published their results first, Leverrier and Galle rightfully got credit from the scientific community for the discovery of Neptune, but history has remembered Adams' role as well.

Historical records have shown that both Uranus and Neptune had actually been observed several times before their credited discoverers found them, Neptune, most notably, by Galileo. He either did not recognize the object for what it was or he was concentrating more on other things he was observing and never took the time to investigate it further.

PLUTO

The discovery of Neptune at 30 AU did not fit the Titus-Bode law. This discrepancy indicates that, even though it did play an important historical role, it is merely an interesting numerical coincidence and not a valid scientific theory. Using perturbations in orbits, as in the discovery of Neptune, was now the preferred method of searching for planets.

Figure 12.5 Lowell

By the beginning of the twentieth century, a search for a "Planet-X" based on small perturbations observed in Neptune's orbit was on. Percival Lowell, a well-to-do businessman who had built an observatory near Flagstaff, Arizona, took the lead in the search. He died in 1916, unsuccessful.

In 1928, straight off his family's farm in Kansas, 18-year-old Clyde Tombaugh asked for a job at the Lowell Observatory. He was initially hired as a maintenance worker, but his proficiency with telescopes eventually got him a job as a full-time staff member. He reopened the search for Planet-X using the observatory's new photographic telescope.

Figure 12.6 Tombaugh

Tombaugh could compare photographs of regions of the sky taken on consecutive nights to look for any object that appeared to be moving in front of the fixed background of stars.

After 10 months of searching, on February 18, 1930, Tombaugh found the elusive Planet-X. He wanted to name the planet Lowell, but keeping with tradition it was called Pluto, the first two letters being the initials of the man who began the search.

Pluto was not at all what was expected. It was much too small to cause perturbations in Neptune's orbit. These calculations were later discovered to be in error, due to uncertainty in the measurements. Before the search for Planet-X, Lowell was best known for thinking he had observed canals on Mars. In 1877 he had heard that Italian astronomer Giovanni Schiaparelli had observed "canali," which is Italian for "channel." Lowell then made an incorrect translation and convinced himself that Martians were building *canals* to move ice from the polar regions of the planet to the equator, so it would melt and supply water for the dying jungles. This scenario gave him a label in the astronomical community as somewhat of a crackpot. His role in the discovery of Pluto gave him (although posthumously) some redemption, but it is interesting that the calculations on which he based his search turned out to be in error. So, had Pluto not been discovered as it was, essentially by accident, the search for Planet-X might have been just as much folly as the Martian canals.

It is also interesting that although Pluto's orbit is very eccentric, its average distance from the Sun, 39.5 AU, is very close to the 38.8 AU "predicted" for Neptune by the Titus-Bode law. Also, the fact that Pluto's orbit actually crosses Neptune's caused some speculation that Pluto was once a moon of Neptune that somehow escaped, causing Neptune's orbit to shift from 30 AU, where it was "supposed" to be. A moon escaping its planet is, physically, very unlikely. What is much more likely is that Pluto and its moon, Charon, discovered by James Christy in 1978, are both members of a population of outer solar system objects known as the Kuiper Belt.

First proposed by Gerard Kuiper in 1951, many small icy objects, also known as trans-Neptunian objects and "Ice Dwarfs," have now been observed beyond the orbit of Neptune. In fact, Neptune may have captured its moons from the Kuiper Belt in a similar fashion to which Mars' moons may have been captured from the asteroid belt between Mars and Jupiter. Over a thousand Kuiper Belt objects are now known to exist, and several similar in size to Pluto's moon Charon, called Quaoar Varuna, and Sedna have recently been discovered.

Activities

1.   Visit a planetarium to observe the current positions and motions of the planets.
2.   Better yet, go outside and see the planets yourself. **Observational Activities** can be found in **Appendix 1**.

Exercises

GRAPHING PLANETARY POSITIONS

Use the figures below (same as Figures 12.1 and 12.2) to fill out the information requested on the tables for Venus and Mars on the next page. Some of the data is given, but what is not you should to be able to figure out. The **Planetary Data** in **Appendix 4** may help.

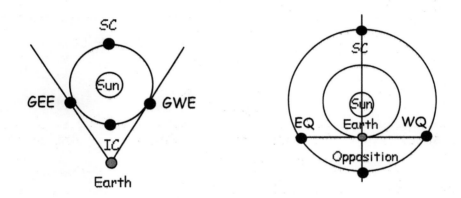

Next, use the tables to plot graphs of the distance from the Earth, the phase and the angle with the Sun for each planet throughout one synodic period. For Venus (the inferior planet) the graphs start at superior conjunction (SC) and go to the next SC, with inferior conjunction (IC) in the middle. GEE and GWE stand for greatest eastern and western elongation, respectively. For Mars (the superior planet), the graphs start at SC, go through opposition (O) and go to the next SC. EQ and WQ stand for eastern and western quadrature.

Venus

Aspect	Distance from Earth (AU)	Phase	Angle with Sun
Inferior Conjunction			
GWE	0.7		
Superior Conjunction			
GEE	0.7		

Mars

Aspect	Distance from Earth (AU)	Phase	Angle with Sun
Opposition			
WQ	1.1		
Superior Conjunction			
EQ	1.1		

Inferior Planet-VENUS

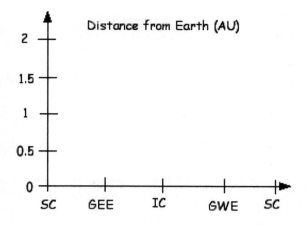

Distance from Earth (AU)

2
1.5
1
0.5
0

SC   GEE   IC   GWE   SC

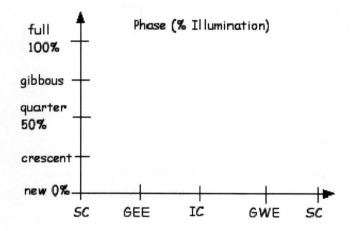

Phase (% Illumination)

full 100%
gibbous
quarter 50%
crescent
new 0%

SC   GEE   IC   GWE   SC

Superior Planet-MARS

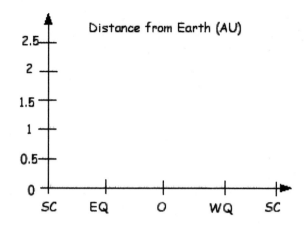

Distance from Earth (AU)

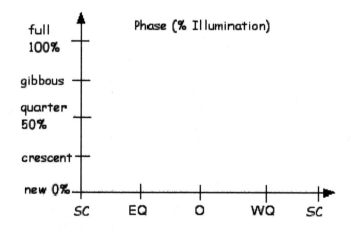

Phase (% Illumination)

110

Inferior Planet-VENUS

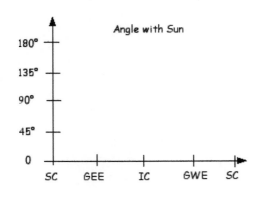

Superior Planet-MARS

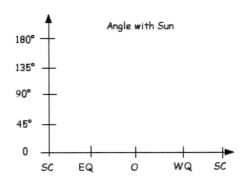

Now it's time to THINK! Try and construct the graphs of how the Apparent Brightness of each planet would change throughout its synodic period. Think about which of the data that you have already plotted should be factors in how bright the planet looks and whether these factors are working together or against each other. There are no numbers on the vertical axes of the graphs. The main concern here is the *shape* of the graph, the graph should be higher at times the planet is brighter and lower when it is dimmer.

Inferior Planet-VENUS

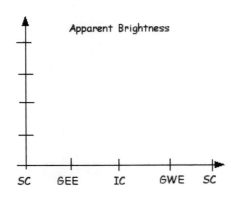

Superior Planet-MARS

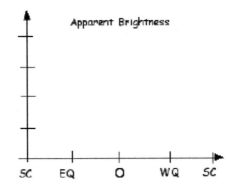

Questions

1. Use Figure 12.1 to determine the phase of an inferior planet in each of the following configurations:

   inferior conjunction- New

   superior conjunction- full

   greatest elongation- Gibbious

2. In which configuration in Figure 12.1 will an inferior planet be visible just before sunrise? Just before sunset?

   Sunrise - Greatest Western elongation
   Sunset - Greatest Eastern elongation

3. Use Figure 12.2 to determine the phase of a superior planet in each of the following configurations:

   superior conjunction-

   opposition-

   western quadrature-

   eastern quadrature-

4. Assuming the Earth rotates counterclockwise, use Figure 12.2 to determine what time a superior planet rises, transits, and sets in each of the following configurations:

superior conjunction-

opposition-

western quadrature-

eastern quadrature-

5. Venus is 0.7 AU from the Sun, and Earth is 1 AU from the Sun. Determine the distance between Venus and the Earth in each of the following configurations:

Inferior Conjunction-

Superior Conjunction-

Hint—<u>Use Figure 12.1 for help</u>

6. Mercury is 0.4 AU from the Sun. Answer question 5 for Mercury.

7. Mars is 1.5 AU from the Sun, and Earth is 1 AU from the Sun. Determine the distance between Mars and the Earth in each of the following configurations:

superior conjunction-

opposition-

Hint—Use Figure 12.2 for help

8. In which configuration in question 7 is Mars furthest from the Earth? How many times further away is Mars when it is at its farthest from the Earth than when it is at its closest?

9. Jupiter is 5 AU from the Sun. Answer questions 7 & 8 for Jupiter.

10. Use the Titus-Bode law (see Figure 12.3) to determine a position for a current Planet-X, a possible next planet after Pluto.

A slightly better image of Uranus than Herschel saw. Taken at Edinboro University of Pennsylvania Maize Sunfire Observatory. Used with Permission of the Director.

Chapter Thirteen

EXTRA-SOLAR PLANETS

According to theories on stellar evolution, or the lives of stars, planets should be common. However, since planets are so much smaller than stars and shine only from light reflected from the stars, visual detection of planets, even those orbiting the closest stars, which are still tens of trillions of miles away, is a virtual impossibility. In the past, astronomers often claimed to have found evidence of a planet orbiting a distance star, but later it would usually be shown not to be the case.

That all changed in 1995 when astronomers found evidence of a planet orbiting the star 51-Pegasi using an indirect method of detection. The method makes use of the fact that, in reality, one object does not orbit another, rather the two objects orbit the center of mass between them. The center of mass of a system will be closer to the more massive object. It will be as many times closer as the number of times more massive the object is than the other. For example, Pluto is about eight times as massive as its moon Charon, so both objects orbit a point that is one-ninth the way from Pluto to Charon, a point eight times closer to Pluto.

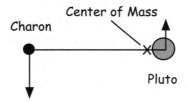

Figure 13.1 Pluto and Charon orbiting the center of mass of their system

In the case of a star and even a very large planet, the center of mass will be so much closer to the star, that the star will move in a path much smaller than the planet's orbit that has the same period as the planet. This motion will be detectable through variations in the star's light. The light will be redshifted when the star is moving away from the observer in its circular path and leaving its waves behind, causing them to spread out. When the star is moving towards the observer, it will catch up with its waves and bunch them up to cause a shift toward the blue end of the spectrum. Not surprisingly, this is called a blueshift.

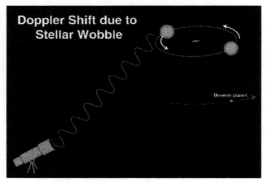

Figure 13.2 Redshifts and blueshifts in a star's light as it moves away from and then towards the observer.

The amount of time for the star's light to go through a cycle of repeated redshifts and blueshifts is the period of the star's motion and is also the period of the planet. Knowing the orbital period, Kepler's third law can then be used to determine the planet's orbital distance. The maximum amount of redshift or blueshift can tell us the speed of the star's motion. Knowing this speed allows the distance the star is from the center of mass to be determined. Finally, using the mass of the star (usually already known from its brightness) and the planet's orbital distance allows us to determine the mass of the planet. This is actually a lower limit for the planet's mass. This is because we have no way of knowing if we are observing the distant planetary system edge on, in which case our mass would be correct, or at an angle, in which case we would not be able to see the entire shift in the star's light. This would cause us to underestimate the mass.

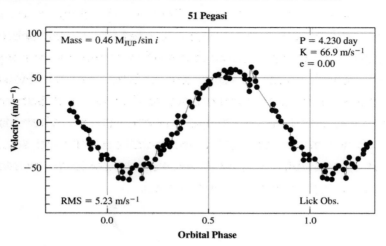

Figure 13.3 Alternating blueshifts and redshifts of the light from the star 51-Pegasi. The period of the unseen planet and the orbital velocity of the star can be used to determine the orbital distance and mass of the planet.

There is an ever-growing list of stars with planets in orbit. Some of them, like Upsilon Andromeda, even have multiple planets.

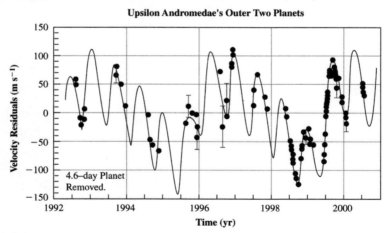

Figure 13.4 Shifts in the light from Upsilon Andromeda, the complex form of the wave indicates the presence of multiple planets, in this case three.

119

The first planets detected were more massive than Jupiter and in orbits very close to their stars, in most cases much closer than would be expected for planets of that size. This is not surprising because the detection method favors more massive planets in closer orbits. The very massive planets cause more noticeable shifts in their star's light that will be repeated rapidly due to a short orbital period. However, more recently, planets of masses comparable to Jupiter and Saturn have been found in similar orbits to Jupiter and Saturn, with no other planets closer to the star. Since these detection methods cannot find smaller planets like Earth, it is possible that smaller planets could be present and the systems are similar to our own solar system.

Exercise

Extra Solar Planets on the Internet

Go to www.exoplanets.org. Once you reach the *California & Carnegie Planet Search* page, click on *General Information*.

On the *General Information* page click on the *Doppler Detection Method*. Read about this method and summarize it.

The wavelengths from a star are measured over a period of time and depending on the measurement is how much the star wobbles which can determine the mass of a planet orbiting that star.

Click on *The Basic Physics Equations used....* The three basic properties that define a planet are the period of its orbit, p, the orbital radius, a, and its mass, m.

Notice that you get the period, p, of the planet from the graph of the Doppler Shift vs. Time. Once you have the period, which of the equations could you use to determine the orbital radius?

$$r^3 = \frac{GM_*}{4\pi^2} p^2$$

Now explain in words how you would use the other two equations to determine the planet's mass, m.

After you get the radius of the planet you can get the velocity of the planet by using $V_{PL} = \sqrt{GM_*/r}$. Then using the velocity you could find the mass of the planet by using the equation $M_{PL} = M_* V_* / V_{PL}$.

Now click on *Public Site* then click on the image of the graph of the known extra-solar planets to enlarge it and click on *Masses and Orbital Radii...* to bring up the table of extra-solar planets.

How many planets are there on the table? -

 110

Which star has the most massive planet? What is its mass? -

HD162020 — 14.4 × M_{jup}

The least massive? What is its mass? -

HD 49674   0.11 × M_{jup}

Which star has the planet in the closest (fastest) orbit? What are the radius and period of this orbit? - $r^3 = \frac{GM}{4\pi^2} P^2$ $r^3 =$

HD73256

P= 2.548

Which star has the planet in the farthest (slowest) orbit? What are the radius and period of this orbit? -

122

Find and click on several stars that have planets in orbits of low eccentricity. Sketch the shape of their Doppler-shift graphs in the space below. Label the graphs with the names of the stars (stay away from stars labeled b or c for now).

Now find and click on several stars that have planets in orbits of high eccentricity. Sketch the shape of their Doppler-shift graphs in the space below and label them (again stay away from stars labeled b or c for now). What differences do you notice from the low-eccentricity orbits?

Find _51-Pegasi_, the first _exoplanet_ discovered (1995). The orbital velocity of the star, K, can be calculated from the following equation derived from the formulas you looked at above.

$$K = 2\pi \, (m/M)(a/p)$$

M is the mass of the star, m of the planet, a and p are the orbital radius and period of the planet. Adjusted for the units used in the table, the equation is of the form K=10396(m/M)(a/p). Use this form of the equation and information from the table to calculate the value of K for 51-Pegasi. Is it the same as the K value given in the table? Note, you will have to click on _51-Pegasi_ and go to its page to find the exact mass of the star.

Find the stars <u>HD 46375</u>, <u>HD 16141</u>, and <u>HD 195019</u> and record their information on the table below.

Star	m (Jupiters)	p (days)	a (AU)	K (m/s)
<u>HD 46375</u>				
<u>HD 16141</u>				
<u>HD 195019</u>				

Pick which graph below is most likely to represent the Doppler-shift data of each of the three stars on your table. Look at the values of m, p, a, and K you recorded and remember the equation $K = 2\pi \, (m/M)(a/p)$. In the space next to each graph write the name of each star and explain why you picked it.

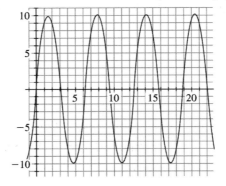

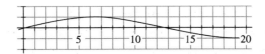

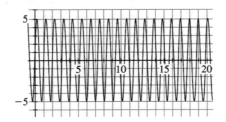

124

Now go to *exoplanets.org/mult.shtml* . Click on several multiple planet systems and look at the form of their Doppler-shift graphs.

In general, how do these graphs seem to differ from those for stars with single planets? -

Single planets will result in Doppler-shift variations that repeat in the form of a simple sine wave (like 51-Pegasi), but when more than one planet is present the combination of two sine waves gives a more complicated waveform.

Click on *47Uma* and compare the data for the two planets.

Which planet has a longer period? b / c (circle one)

By about how many times?

Which planet has a greater K?   b / c (circle one)

By about how many times?

Circle the graph below that you think is likely to be for planet b (the other will be for Planet c).

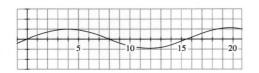
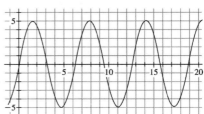

125

Now combine these two graphs to reproduce the graph for 47Uma. Find and open the program *Graphing Calculator* (or use a similar application) on your computer.

Type in the equation;

$$Y = 5\sin x + \sin(x/2.5)$$

Each term in the equation is for one of the planets. Look at the graphs above.

Which term is for planet b?   y=

Which is for planet c?   y=

Sketch the graph you made in the space below. Does it look similar to the actual graph for 47Uma?

Return to _exoplanets.org/exoplanets_pub.html_ and select one of the **What's New** stories that interests you. Read it and summarize it in the space below.

Epilog

THE MUSIC OF THE NEW SPHERES?

The star HD 82943, see Figure E.1, has two planets with a ratio of orbital periods very near 2:1. This yields a waveform very similar to that of a musical *octave*. Two musical pitches that are an interval of an octave apart also have a frequency ratio of 2:1. The interval between two musical notes of the same name, like *C-C'*, the first and last notes of a musical scale, is an octave. Figure E.2 shows the waveform produced by two tuning forks of frequencies *C-256 Hz* and *C'-512 Hz* sounding simultaneously. Note the similarity to the waveform for HD 82943 in Figure E.1.

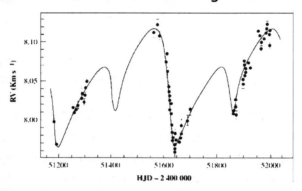

Figure E.1 HD 82943 has two planets that orbit with periods in a ratio of 2:1.

Could this be an example of "The Music of the NEW Spheres"? In the 6th century BC, Pythagoras discovered that portions of a plucked string that correspond to simple fractions of the entire string, i.e. 1/2, 2/3, 3/4, etc., sounded pitches that formed consonant or "harmonious" musical intervals with the pitch sounded by the entire string. Since that time, the search for simple numerical relationships, or harmonies, in nature has been an undercurrent of much scientific investigation. It is well known that Kepler believed in "Cosmic Harmony." As mentioned in **Chapter 8**, this belief led him to conclude that each planet "sung" as it orbited the Sun. Kepler discovered his third or "Harmonic" law as a result of searching for a simple or "harmonious" relationship between the quantities that describe a planetary orbit, a search that took him ten years. This relationship, the square of a planet's period being equal to the cube of its average distance from the Sun, is how we determine the orbital radius of the extra-solar planets once their period is known from observation of the varying Doppler-shifts.

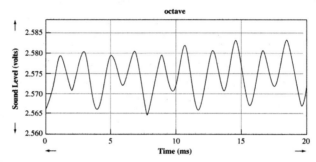

Figure E.2 The waveform produced by simultaneously sounding two tuning forks having a frequency ratio of 2:1 (a musical octave).

There have been other examples of "Musical Spheres" or "Cosmic Harmony" since the time of Kepler. For instance, the pitch of sound is analogous to the color of light; they are both caused by the frequencies of their waves. Because of this, the ratios between the frequencies of the four visible lines in the Hydrogen spectrum have been compared to the intervals in a musical chord. Hydrogen makes a dissonant chord, but a chord nevertheless. There is no doubt that any harmonies found in the Cosmos would have interested Kepler, but it is likely that this music of these new spheres that are being found, partially through the use of the laws he discovered so long ago, would have been among his favorites.

Appendix One

STAR OBSERVATIONS

Daily Motion: Go outside on a clear night about an hour after sunset. Find the Big Dipper and use it to find the North Star, Polaris. To review how to do this, see Figure 1.4. In the space below on the left, draw the Big Dipper and the North Star. Now wait three hours and do the same thing in the space on the right. Note any differences you see.

Big Dipper and North Star   Big Dipper and North Star

Date-   Time-   Date-   Time-

What differences do you notice?

131

STAR OBSERVATIONS

Annual Motion: Go outside on a clear night about an hour after sunset and look in the southern sky. In the space below draw a constellation that you can see. Use the seasonal star maps in **Appendix 2** for help identifying constellations. Do this activity once near the beginning of the semester and once near the end.

Constellation in South Sky

Date-   Time-

Constellation is South Sky

Date-   Time-

SUN OBSERVATIONS

Observe the Sun four times throughout the semester, preferably spaced out about once per month. Choose either sunrise or sunset and make all four observations at this time. They must also all be made from the same location. In the spaces on the **Sunrise/Sunset Observations** page, draw a picture of the horizon and any nearby landmarks (trees, buildings, etc.) and draw the Sun in its position on the horizon. The only thing that should be different in the four drawings is the location of the Sun.

MOON OBSERVATIONS

On every clear night for two weeks, beginning with the date of the New Moon, go outside, face due south, and find the Moon. In the spaces provided on the **Moon Observations** page, record the date and time of four of the observations during the two weeks and draw the Moon, showing its shape and its position relative to the Sun. Try to space out the observations you record as evenly as you can over the two weeks. You should be facing south for all your observations, so east would be on the left of the drawing and the setting Sun in the West on the right. A new Moon occurs about once per month, so you should have at least three and maybe four chances to complete this exercise in a semester.

Sunrise / Sunset (circle one) Observations

Date of Sun Observation-

Date of Sun Observation-

Date of Sun Observation-

Date of Sun Observation-

135

Moon Observations

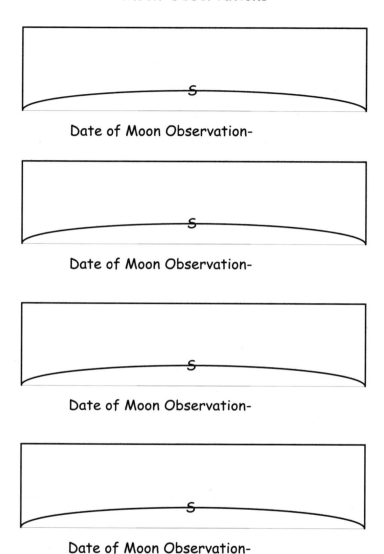

Date of Moon Observation-

Date of Moon Observation-

Date of Moon Observation-

Date of Moon Observation-

PLANETARY OBSERVATIONS

As you have learned, each planet has a unique sidereal period, so their positions in the sky will vary from year to year. Their visibility will also vary as they move through their synodic period. Nevertheless, when it comes to observing planets there is almost always something interesting to see. Your instructor can assign you what to look for as well as where and when. What you see can be drawn in the space below. Also, monthly publications such as *Astronomy* or *Sky and Telescope* always keep track of planetary positions as well as many other astronomical events.

Appendix Two

SEASONAL
STAR MAPS

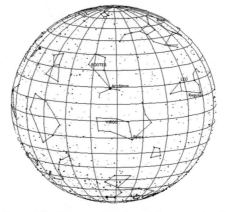

(Looking south at mid-evening)
Spring

(Looking south at mid-evening)
Summer

139

(Looking south at mid-evening)
Fall

(Looking south at mid-evening)
Winter

140

Appendix Three

CELESTIAL GLOBE ACTIVITIES

* Note—These activities were designed for the Celestial Globes used at HFCC, but could be adapted to apply to any number of similar models.

Celestial Globe Activity #1-STARS

1. Set the Celestial Globe for your Latitude. You do this by moving the top of the globe (90° Declination), the North Celestial Pole, toward the North so that the Latitude you want is at the Zenith, the point directly overhead. Your Latitude =_____°.

   a. Find the four Constellations in the table below. Classify them as stars that either: rise and set (R/S), never set (NS), or never rise (NR) at your Latitude. Also give the name of a bright star in the Constellation (if one is labeled).

Constellation	R/S	NS	NR	Bright Star
CANIS MAJOR				
Cassiopeia				
Crux				
Cygnus				

b. Leave the globe set at your Latitude. Determine whether the Transit Altitude (the Altitude at which they cross the Meridian) of the stars named in your above table are high-h, medium-m or low-l. High would be close to the zenith, low would be close to the horizon. Also, determine the Azimuths on the Horizon when they rise and set. Are they rising in the NE, E, or SE? Are they setting in the NW, W or SW?

Star Name	Transit Altitude (h/m/l?)	Rise Azimuth (NE,E,SE?)	Set Azimuth (NW,W,SW?)

2. Use the Celestial Globe to answer the following questions:
   a. Is the Altitude of the North Celestial Pole as seen from your Latitude

   h / m / l (circle one)?

   b. Now set your Globe to 90° N Latitude. Your geographic location is (where on the Earth are you observing from?) _____. The Altitude of the NCP is

   h / m / l (circle one).

142

c. Find the same four Constellations you found in 1(a). In the table below, classify them as stars that either: rise and set (R/S), never set (NS), or never rise (NR) from 90° N Latitude (a constellation can only be ONE of these choices from a given location). Also, transfer your results from part 1(a) to complete the third column.

Constellation	90°N	Your Latitude = _____ °	0°	90°S
CANIS MAJOR				
Cassiopeia				
Crux				
Cygnus				

d. Now set your Globe to 0° Latitude. Your geographic location is (where on the Earth are you observing from?) _____. The Altitude of the NCP is h / m / l (circle one). In the table above, classify the four Constellations as stars that either; rise and set (R/S), never set (NS), or never rise (NR) from 90° N Latitude (again, a constellation can only be ONE of these choices from a given location).

e. Compare the Altitude of the NCP in parts 2-a, 2-b, and 2-d. How does the Altitude of the NCP seem to be related to the observer's Latitude?

f.   Now set your Celestial Globe to 90° S Latitude. Your geographic location is (where on the Earth are you observing from?) _____. In the table above, classify the four Constellations as stars that either: rise and set (R/S), never set (NS), or never rise (NR) from 90° S Latitude. Can you see the NCP from here?

g.   Of the locations you visited, which has the most never-set (or never-rise) stars? Which location has the least never-set or never-rise stars?

h.   Which location that you visited has the most rise and set stars?

     Which location has the least rise and set stars?

i.   About what percentage of the stars visible from your location are rise and set stars?

144

j.   As you travel from the equator to a pole the number or never-rise and never-set stars

increases/decreases   (circle one)

and the number of rise and set stars

increases/decreases   (circle one).

*   After you have completed this activity, you should be able to;

1.   Set the Celestial Globe at any Latitude.
2.   Determine the approximate Altitude (h /m /l) and Azimuth (N, NW, SW etc.) of an object.

145

Celestial Globe Activity #2-THE SUN AND THE SEASONS

1. Set the Celestial Globe for your Latitude. You do this by moving the top of the globe (90° Declination), the North Celestial Pole, toward the North so that the Latitude you want is at the Zenith, the point directly overhead. Your Latitude =_____°.

   Now move your Sun to the current date on the Ecliptic.

2. The Sun is in the constellation _____ .

   The Sun rises in the   NE / E / SE (circle one)

   The Transit Altitude is h / m / l ?

   What time of day is this? -

   The sun sets in the   NW / W/ SW   (circle one)

3. Name a Constellation that will be visible high overhead:

   shortly after sunset-

   around midnight-

   Explain how you can tell it is midnight-

   shortly before sunrise-

4. Record on the table below the rise and set Azimuth (example-NE, E or SE) for the Sun, the Transit Altitudes (h / m / l) and the approximate number of hours the Sun is up. Do this for each date given.

Date	Rise/Set Azimuths (NE, E, SE etc)	Transit Altitude (h / m / l)	Hours Up**
March 21			
June 21			
September 21			
December 21			

**   **Note-**You can count the <u>HOURS UP</u> by counting how many lines of Right Ascension pass under the Meridian during the time the Sun is up.

Based on the above table, give two reasons that summer is warmer than winter.

March 21 and September 21 are called the Vernal and Autumnal _____s. Based on the table above explain the reason for this name.

5. Classify each Constellation as either seasonal or circumpolar by placing a check mark in the appropriate box. A Constellation will be named after the season you can see it best. This is the season that it is high at midnight. So, put the Constellation on the Meridian then move the Sun to midnight and see what date on the Ecliptic the Sun is on. The season that date is in is the season you will best see the constellation. Circumpolar Constellations can be seen all year around, so they are not seasonal.

	Cepheus	Leo	Lyra	Pisces	Orion
Spring					
Summer					
Fall					
Winter					
Circumpolar					

* After you have completed this activity, you should be able to:
1. Correctly place the Sun on the Ecliptic for any day of the year.
2. Determine the approximate number of hours an object is up.
3. Determine the season in which a Constellation will be best seen.

Celestial Globe Activity #3-AROUND THE WORLD

1. To set the Celestial Globe for any Latitude, move the top of the globe (90° Declination), the North Celestial Pole, toward the North so that the Latitude you want is at the Zenith, the point directly overhead.

   Place your Sun under each of the dates on the Ecliptic in the tables below and record the information for each of the given Latitudes.

TRANSIT ALTITUDE
(h / m / l ?)

Date	40°N	80°N	20°N	40°S
March 21				
June 21				
Sept 21				
December 21				

HOURS UP**

Date	40°N	80°N	20°N	40°S
March 21				
June 21				
Sept 21				
December 21				

** **Note**—You can count <u>HOURS UP</u>, by counting how many lines of Right Ascension pass under the Meridian during the time the Sun is up.

RISE AND SET AZIMUTHS
(E, NE, SE; W, NW, SW)

Date	40°N	80°N	20°N	40°S
March 21				
June 21				
September 21				
December 21				

2. Answer the following questions based on the tables above.
    Which Latitude has the greatest variation in <u>HOURS UP</u>?

    The least variation?

    Was the Sun ever (close to) directly overhead (near the Zenith)?
    If so where and when?

    Did you ever observe "Midnight Sun" (the Sun up all night or close
    to it)? If so where and when?

    Did you ever observe a day with (little or) no sun at all? If so
    where and when?

Explain what reasons **you observed** for the Polar Regions of the world always being so cold and the tropical regions being so warm.

Based on what **you observed**, does the length of time the sun is up seem to be a primary or a secondary effect on both latitudinal and seasonal variations in temperature? Explain your answer. Hint-compare <u>HOURS UP</u> and <u>TRANSIT ALTITUDES</u> at 20°N and 80°N.

*   **This Activity introduces no new skills with the Celestial Globe, but reviews almost all of those previous learned.**

Celestial Globe Activity #4-THE MOON

1. To set the Celestial Globe for any Latitude, move the top of the globe (90° Declination), the North Celestial Pole, toward the North so that the Latitude you want is at the Zenith, the point directly overhead. <u>Set your globe to the Equator.</u>

2. Now place the Sun on the Ecliptic on the date of March 21 (along the 0 hour of Right Ascension). A New Moon occurs when the Moon is at the same Right Ascension as the Sun. Place your Moon on the same location on the Ecliptic as the Sun. Record the Date and the Rising, Setting, and Transit times of the Moon on the table below.

3. The Moon will move completely around the Ecliptic in about one Month.
   This is 24 hours of Right Ascension. So, the Moon will move about _____ hrs of RA in one week. The Sun will move completely around the Ecliptic in one year, so in one month it will move _____ hrs of RA and in one week it will move _____ hrs.
   Now advance the Sun one week along the Ecliptic and advance the Moon one week. The Moon is always within 5° of the Ecliptic, so for this activity it can be considered to move along the Ecliptic too. Record the approximate Date and the Rising, Setting, and Transit times of the Moon on the table and the name of the Lunar Phase. Hint for the Phase-What fraction of the way around the Ecliptic has the Moon moved since New Moon?

4. Now advance the Sun another week and the Moon another week. Record the Date and the Rising, Setting and Transit times of the Moon on the table and the name of the Lunar Phase. Hint-Is the side of the Moon facing the Earth, facing toward or away from the Sun?

5. Advance the Sun and the Moon one more week. Record the Date and the Rising, Setting, and Transit times of the Moon on the table and the name of the Lunar Phase. Hint-What fraction of the way around the Ecliptic has the Moon moved since New Moon?

6. Finally, advance the Sun and the Moon one more week. Notice that the Moon has NOT returned to the New Phase. The Moon is about _____ hrs of RA "behind" the Sun. It will take about _____ days to "catch up" and return to the New Phase (divide the number of hours of RA in a week by the number of hours of RA the Moon is behind, then divide the answer into the 7 days in a week).

The time for the Moon to move all the way around the Ecliptic is called the _____ month and is about _____ days. The time between two New Moons is about _____ days and is called the _____ month.

Moon Table

Record the Phase, the Date and the Rising, Setting and Transit times of the Moon from questions 2-5 above. Enter the times as sunrise, noon, sunset, or midnight—which ever is CLOSEST!

Phase	New			
Date	March 21			
Rise Time				
Transit Time				
Set Time				

156

7. Answer the following questions based on the table above.
    During which phase(s) will you see the Moon mostly at night?

    During which phase(s) will you see the Moon first during the day,
    then at night? First at night, then during the day?

    During which phase(s) is a Lunar Eclipse most likely? Explain your
    reasoning and what happens during a Lunar Eclipse.

    During which phase is a Solar Eclipse most likely? Explain your
    reasoning and what happens during a Solar Eclipse.

    During which phase(s) will the Moon not be visible? Explain your
    reasoning.

8. Redo the table for YOUR LATITUDE and begin on line of Right
    Ascension closest to the date of the most recent New Moon. Try to
    explain the reasons for the differences that you see from the first
    table.

Phase	New			
Date				
Rise Time				
Transit Time				
Set Time				

Reasons for differences between the two tables? -

Appendix Four

PLANETARY DATA

Planet	Mass (Earth=1)	Radius (Earth=1)	a Distance (AU)	p Sidereal Period (years)	Synodic Period (days)	e*	i#
MERCURY	.0056	.38	.39	.24	116	0.2	7
Venus	.82	.95	.72	.62	584	0.007	3.3
Earth	1	1	1	1	-	0.017	0
Mars	.11	.53	1.52	1.88	780	0.093	1.9
Jupiter	318	11.3	5.2	11.86	399	0.048	1.3
Saturn	94	9.4	9.6	29.5	378	0.056	2.5
Uranus	14.5	3.98	19.2	84.1	370	0.046	0.8
Neptune	17.2	3.8	30.1	164.8	367	0.01	1.8
Pluto	.0023	.27	39.5	248.6	364	0.248	17.2

*   eccentricity of planet's orbit
\#   angle of inclination of planet's orbit from the ecliptic in degrees

Appendix Five

OTHER BOOKS ABOUT CYCLES IN THE SKY

Michael J. Crowe, *Theories of the World from Antiquity to the Copernican Revolution*, Dover Publications Inc., NY, 1990.

Norman Davidson, *Sky Phenomena-A Guide to Naked-Eye Observation of the Stars*, Lindisfarne Press, Hudson NY, 1993.

James Kaler, *The Ever Changing Sky: A Guide to the Celestial Sphere*, Cambridge University Press, 2002.

H.A Rey, *The Stars-A New Way to See Them*, Houghton-Mifflin Co., Boston, Mass., 1980.

ABOUT THE AUTHOR

Michael C. LoPresto was born May 29, 1965 in Tucson, Arizona. He was raised in Edinboro, Pennsylvania and received a Bachelor's Degree in Physics in 1987 from Edinboro University of Pennsylvania. He also holds an MS in Physics from the University of Michigan, 1989, and an MS in Physics Education, 1996, from Eastern Michigan University.

He has been a member of the Science Division Faculty at Henry Ford Community College in Dearborn, MI since 1990, teaching courses in Physics, Astronomy, and Atmospheric Science. He is currently serving as Chair of the Physics Department and Director of the Planetarium. He has published over 20 articles in journals including the *American Journal of Physics*, *The Physics Teacher*, *Physics Education*, the *Astronomy Education Review*, *The Science Teacher* and *Mercury*. He has made numerous presentations at meetings of the American Association of Physics Teachers, the Astronomical Society of the Pacific, and the Great Lakes Planetarium Association, as well as for the HFCC Cultural Activities Program, the University of Michigan Natural History Museum, and the Detroit Observatory. His research interests include musical acoustics and astrophysics.

Mike resides in Saline, Michigan with his wife, Jan, a musician and music-teacher, their children, Sarah, Emily and Sam and dogs Henry and Schroeder. When not spending time with his family or busy with his job, Mike enjoys music, playing trombone, and singing, following University of Michigan football, reading, and exercise. Also, he has recently become involved in coaching his son's baseball and soccer teams.